Lex Parsimoniae

The Mystery behind
137
Was revealed!

Occam's razor reloaded

By MIGUEL DE ZAYAS

Miguel De Zayas / Occam's razor reloaded / 2

Lex Parsimoniae

"Entia non sunt multiplicanda praeter necessitate"

"THE LAW OF BRIEFNESS"

Occam's Razor Reloaded

Miguel De Zayas

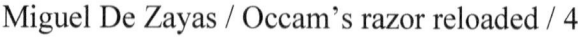

TABLE OF CONTENTS

Introduction

"Entia non sunt multiplicanda praeter necessitate"

What is Ockham's razor and why should we worry about keeping its status up-to-date at all times, especially in the field of science?

Let's start with the first. Ockham's razor is a philosophical principle based entirely on the prevalence of logic. A man named William Ockham, a Franciscan friar who lived in the fourteenth century, proposed what has become ever since a rule of thumb for scientists and philosophers alike "Lex Parsimoniae;" in other words, the "Law of Briefness."

The answer for the second question should become obvious for most people but paradoxically as it may seems, continues to be the subject of many diverse interpretations and controversy.

Allow me to explain what the razor states first:

"The simplest solution is usually the correct one." Eight words representing an objective truth.

Five years ago I published a book titled "Warning Scientific Discretion Advised." In the absence of empirical evidences I had no choice but to publish it in the context of a Sci-Fi assay meant to satisfy my own personal "ego." I confess that back then its final publication brought with it a sign of personal relief. I had the chance to express to the "entire world" (☺) my dissatisfaction in regards to the way science has failed in its prior attempts to provide an objective representation of reality while labeling at the same time as a technical inconvenient the concepts of logic and common sense. I wrote it for the records independently of the fact that in the lapse of time between now and the past five years I have only sold one single piece! My thoughts (For the records) were recorded even if almost no one has ever read it... In fact, if according to a probabilistic theory "a book only exists if someone actually read it" I had no choice but to arrive to the sad conclusion that perhaps my first book never actually existed in the first place.

To provide you with a short glimpse of how upset I felt about the manner in which logic and common sense was handled in the twentieth century I selected a couple of examples below.

The following was a quote from a book titled "The Concept of Science" and appears in page **45, 8**[th]. Paragraph of my book "Warning Scientific Discretion Advised":

"... Generally when a person does not accept a theory, it is not because he does not like theories but because that particular theory does not agree with its own prejudices... which he likes to describe as common sense. If there is one thing that the developments of 20th

century physics have taught us, it is that we cannot trust common
sense... as a criterion for the truth or falsity of a physical theory..."
[End of quote].

The above paragraph obviously refers to an increasingly broad
spectrum of philosophical conclusions derived from the so-called
"Copenhagen Interpretation" while reserving a pivotal role to the
observation as well as the "observer."

To answer those attacks aimed at discrediting the legitimacy of
logic and common sense as tools in our eternal search for the
"final truth", I wrote the following on page **46, 4**th Paragraph:

**"Common sense is more than just a philosophical basis for scientific
inquires aimed to answer Socrates questions. It is more than a
political commodity conveniently used at our own discretion, and it
is definitely more than just an undesirable obstacle that happens to
get in the way between our judgments and those "new theories" of
yours..." (Namely Quantum Mechanics and General Relativity)...
"Common sense is a gift from God!"** [End of quote].

A gift (especially coming from our creator) is in my humble
opinion something we must always keep, celebrate and cherish
whether sometimes we might feel uncomfortable with it or not.
Ockham's razor was meant precisely to remind us about the
infinite value of such a marvelous gift. The day we began to
question the legitimacy of its own existence we lost our way

inside "wheeler's tunnel." I believed it was John Archibald Wheeler who said that "worse than a tunnel that goes on forever is a tunnel that has an end".

This book was written with the sole aim of sparking a light at the end of that tunnel... If I achieve my goals, it will spread into every single field of science as a fire sparked by a bolt of lightning does across the forest.

I heard somewhere someone saying that "Stream circumstances sometimes require the use of steam measures" so, I will demonstrate the beauty, simplicity and universal application of a theory relying solely (as it should always be) on a body of purely empirical evidences.

As many of you may know from the annals of scientific literature, Albert Einstein owed most of his findings and remarkable conclusions to the use of "thought experiments." In a thought experiment you rely only on two things: the power of your imagination and the ability of abstraction. I must admit that the completion of this seemingly absurd enterprise required from me a sacrifice of extenuating hours of search among a long list of patterns and their analysis in the course of few years until finally I was able to visualize what it began with a crazy hypothesis of mine.

Many of you may have heard an old saying that **"God works in mysterious ways"** and nothing could offer a better proof of that than the amazing results that I achieved in the last few days. I will attempt to introduce you to the fundaments of a completely new theory that will revolutionize the world in ways no one could even imagine today. A new concept of what I see as our objective reality, the same one Albert Einstein was trying to figure out in his thoughts and the only one that would enable humanity to offer a complete description of every single experiment along with its expected predictions. My theory will hopefully offer the needed answer to enigmas like the mystery behind the "Double Slit Experiment", "Quantum Entanglement", "Molecular interactions" and the true nature of gravity among many other subjects not to mention the benefits that will bring in the field of genetics particularly in the understanding of the mechanics behind the mutual interaction between genes.

What do we have so far? Well, it seems that all hopes have been placed on a theory notoriously complex with records of zero-predictive powers whatsoever named 'String Theory." The following is a remarkable quote that echoes my own views as well:

"For more than a generation, physicists have been chasing a will-o'-the-wisp called string theory. The beginning of this chase marked

the end of what had been three-quarters of a century of progress. Dozens of string-theory conferences have been held, hundreds of new Ph.D.s have been minted, and thousands of papers have been written. Yet, for all this activity, not a single new testable prediction has been made, not a single theoretical puzzle has been solved. In fact, there is no theory so far—just a set of hunches and calculations suggesting that a theory might exist. And, even if it does, this theory will come in such a bewildering number of versions that it will be of no practical use: a Theory of Nothing." (End of quote)

And that's not all! What about the concept that conceptualizes the title of this book? How about the notion that a good theory must be built upon three major principles: simplicity, briefness and predictability?

I'll demonstrate in this little book with nothing but words coming out from my fingertips that "String Theory" has been nothing but a waste of resources and what's even more tragic, a waste of time.

At this point I'd like to quote one of my favorite physicists of all times, Dr. Wolfgang Pauli. These words speak for themselves:

Das ist nicht nur nicht richtig, es ist nicht einmal falsch!

"Not only is it not right, it's not even wrong!"

I will end a brief introduction to my theory quoting once again a classic paragraph that I also inserted in page 17, paragraph 2nd of my book "Warning Scientific Discretion Advised" and reads as follows:

(Fragment of a letter Albert Einstein wrote to his dear friend Max Born)

"We have become Antipodean in our scientific expectations. You believe in the God who plays dice and I in complete law and order in a world which objectively exists, and which I, in a widely speculative way, am trying to capture. I firmly believe, but I hope that someone will discover a more realistic way, or rather a more tangible basis that it has been my lot to find..." [End of quote].

Einstein was extremely concerned about the probabilistic interpretation given to the model of our reality by Quantum Mechanics and its followers. He kept his all life looking for a new breed of numbers, a completely new math that would change the world. As I posted in my first book in regards to the roll of mathematics in science, I said that it represents the language of science and it's nonetheless an invaluable tool during a calculation, but I also wrote in my first book that numbers alone were not oracles.

One can't blame others for not been able to see what only a single soul could, however it was a sin trying to impose a view on someone under the weight implied by the phrase: **"Shut up and calculate!"**

After I finally succeeded in publishing my first book I realized that I was back on 'square one'. I felt good in the inside but I ended up with no specific theory and what's even worse than that, I couldn't prove a single argument I spent writing for almost an entire year. I had to find out and be able to come up with a strong body of evidences without relying on any sort of sophisticated laboratory facilities (That I didn't have access to begin with). I literally speaking needed a "miracle;" something that would unanimously convince everyone that I was "up to something," not because I was eager to be finally 'considered' or as a 'revenge' for being ignored in the past, but because I knew from the bottom of my heart that I was right.

Was it a miracle having understood the process hidden inside the very 'fabric' of our Reality?

I could say that this question hasn't been entirely alien to me... especially these last days.

Did you know that there was a very famous professor who said that miracles do happen to people?

J. E. Littlewood, a Cambridge professor, defined a miracle as an exceptional **event** of special significance occurring at a frequency of one in a million... I'm not joking! His theories have been successfully published and can be found inside a collection of his work under the title "A Mathematician's Miscellany".

Am I solid evidence in favor of Dr. Littlewood's theories or was it something else?

I knew, just as almost everyone involved in the search for a unified theory or at least for the basic principles capable of sustaining such a claim, that whatever is out there must be acting unconditionally on both worlds: the micro-world and the macro-world at the same time. Since the micro-world (quantum realm) was observable only under special laboratory conditions I came to the conclusion that perhaps looking on the other direction I might get **lucky** and could finally "bring some order to the chaos."

After a long and seemingly endless search for the meaning of strange patterns that fascinated me for years, I could say that I finally have some empirical evidence to present to you (the reader) about an entirely new model of reality and one that I visualized and attempted to describe more than five years ago on pure intuition. Neither String Theory nor Quantum Mechanics or any other among the long list that claims the existence of many worlds and a virtual universal wave (Bohm's interpretation) has so far been able to PREDICT where the "unpredictable" electron was going to hit once it reaches the screen of the "Double Slit

Experiment". These are the facts and they remain until today undisputable ones (!)

Until today, no one has even come up close to explain WHY our DNA manages to mutate and shuffle its genes during the evolutionary process of existence.

Ladies and Gentlemen understanding Reality means being able to make a simple connection among all I have said and what's even more exciting yet: being able to PROVE IT.

I can understand that the "Law of Probabilities" has been a historical landmark of how well the philosophical tenet contained within the Occam's razor has served humanity in the search for better and simpler ways to represent Reality... but that will end right here and right now!

I'll prove that the "Law of Probabilities" has been nothing more than a primitive attempt to represent an incomplete face of our Reality in a pitiful two dimensional context.

CHAPTER I

"R E A L I T Y"

The following quote was expressed by Dr. Jesse L. Greenstein of the California Institute of Technology:

"The detection of gravitational waves bears directly on the question of whether there is any such thing as a "gravitational field," which can act as an independent entity. … This fundamental field hypothesis has been generally accepted without observational support. Such credulity among scientists occurs only in relation to the deepest and most fundamental hypotheses for which they lack the facility to think differently in a comparably detailed and consistent way. In the nineteenth century a similar attitude led to a general acceptance of the ether..." (End of the quote).

I could add yet another example to his clever remarks by mentioning the same behavior regarding the nature of the "magnetic field" after Maxwell's equations were published.

The three equations trying to represent the properties of space inside the boundaries of a "spatial band" filled with some unknown "electromagnetic fluid" led most physicists to accept the concept of a magnetic field back then too.

As you're reading this page, a heated controversy regarding **the speed of gravity** is going on; whether it is the same as **c** (the speed of light) or perhaps other than **c.** The obvious question derived from the first would undoubtedly follow up with another question that constitutes the key in the understanding of Reality in accordance with my theory.

Is the speed of light a constant throughout the universe the way Albert Einstein postulated in his General Theory of Relativity or perhaps was his decision to "sacrifice" the topography of space-time nothing but a huge blunder?

Unfortunately I must admit today in this book that he made a great mistake unintentionally bringing everyone else with him deeper into "Wheeler's Tunnel."

Before I answer this question I decided to illustrate with real names some of the 'players' involved in the latest contest:

One among this group is a physicist called Steve Carlip. He stated that the planetary orbits calculated by Einstein's General Relativity are the same as those derived from Newton's mechanics "If" we were to admit "instantaneous action at a distance."

Let's make clear what he just said. He compared Einstein's calculations and theory with a much older one (which by the way works perfectly even today) assuming the author of General Relativity would have been willing to **give up** a stubborn attachment toward a constant and universal speed of light. Too late! Besides, why do you think, Mr. Carlip that Einstein was so depressingly concerned after the final conclusions concerning entanglement and the violations of Bell's inequalities? Unless you accept and understand my arguments about the nature of Reality the only logical explanation to the experiments concerning entanglement was the legitimacy of a "force acting in both places "Alice" and "Bob" instantaneously. (!)

We all know Sr. that **E.P.R.** was a disaster both in design and predictive power and not precisely because Quantum Mechanics was right (The way was concluded by some) but because it failed short to provide us with the real representation of the fabric of Reality.

I'd like to remind you that we are talking about gravity and the Solar System and not about quantum entanglement... However, have you noticed a coincidental 'overlap'?

Well there is nothing left to accidental in all this and my theory proves it.

Clifford M.Will and Edward Formalont claimed that the speed of gravity fluctuated between **0.8** and **1.2** times **c** after a series of observations done on the planet Jupiter.

A Japanese physicist named Hideki Asada immediately rushed to condemn their hypothesis accusing them as another futile attempt to revive an old controversy about the actual values of the speed of light. The whole issue fell into the well know "chicken and the egg" paradox... Hasn't always so?

This is literally speaking, a quote from Wikipedia:

"It is important to understand that none of the participants in this controversy are claiming that general relativity is "wrong". Rather, the debate concerns whether or not Kopeikin and Fomalont have really provided yet another verification of one of its fundamental predictions." (End of quote)

It all comes to the same point at the beginning of this chapter: Reality.

I mentioned that my first book had to be published pretending to be a sci-fi assay rather than a serious attempt to confront those issues face to face. I also explained why. I didn't have any empirical evidence to prove my theory in the first place.

In more than one chapter of that "book of fiction" I wrote that "ALIENS" had based their theory of gravity (Temporal Mechanics) on a universal principle that reads:

"In every single possible event where **Symmetry** could be violated, the universe will respond by trading space for symmetry." I used the same theoretical principle to fundament the working principles behind "Time Thrusters" in a chapter that holds the same title.

How could you translate this simple principle to the question whether "instantaneous action at a distance" may or may not be possible?

Both entanglement and those 'inconsistencies' we can't understand are vivid examples of the instantaneous application of the same "Alien" principle stated in a science fiction novel.

The speed of light is not in any way "immune" to the universe's response when it comes to preserve the integrity of any of the 3 most important principles ruling our Reality; "The Law of Conservation of Energy", "The Law of Conservation of Symmetry" and the "Conservation of Total Perturbation".

The deciding factor in favor of General Relativity against Newton's Mechanics was the observational errors regarding the precession of Mercury around the Sun calculated upon incomplete theories and mistaken principles. We allowed the abstract concept of a curved space-time continuum to go on because we couldn't understand the **'missing link'** connecting "Quantum realm" with "Celestial Mechanics".

Remember that for Einstein the discovery of this "link" was crucial in the final representation of an objective reality. He tried to find it until his time ran out but unfortunately he fell short.

Keppler was the first one who stated the universal concept for a planetary symmetry both in words and in the language of science (Math) when he so wisely argued that our planet Earth had to cover equal areas at all times during its voyage around the Sun in the name of symmetry preservation. No one dares today to contradict such a well proved fact, right?

Couldn't this sudden acceleration be considered a simple evidence of how **Reality** intervenes in a possible case where symmetry becomes 'threaten' by the shape of an elliptical orbit? **The scales are different** and the laws ruling both worlds too but the same behavior is present in a different context as it is the case with Fourier series involving heat, sound and electromagnetism... An observer a few light-years from the Solar system would probably feel about Earth's sudden acceleration toward the Sun during the precession of the equinoxes just as weird and frustrated as we do today about planet Mercury from its "door-step".

Let's be clear once and for all about the notion of the speed of light in the vacuum. When the Ether was finally dimmed obsolete and a more modern word came to replace it (Vacuum) we thought that we had finally turned the page over with regards to the chapter of space and time. Einstein contradicted himself in more than one occasion when asked about the (worrying) emptiness implied in his newly proposed 'vacuum.' If indeed the vacuum was just empty space then, what medium was used to guide radio waves across two distant points?

Some physicists came up with a wave configuration that appeared to be self-sustained and in no need for a medium... Funny isn't it?

The medium has always been there. The problem that it was "invisible" (until proved otherwise) led to a logical assumption that extends the rest of the universe with the same vacuum; a unique fabric of space-time "tailored to fit" every single region of the known universe. We just needed to learn how to bend it and warp it to our convenience until we match prediction with observation...

In our Solar System the vacuum is a multilayer harmonic configuration of spherical "Standing Waves" built upon orbits and making up waveguides for gravity; not space and time, but Reality was the only responsible for each planetary momentum around the Sun.

There is just one question left out from our conversation... How is the universe "capable" of guarding and protecting the conservation of symmetry?

What is the physical explanation behind an inexplicable acceleration in order to keep symmetry intact? What is causing the eerie results observed by photons observed at "Alice" after "Bob" suffered from a wave-collapse?

The answer lies in the particular structure of the harmonic spectrum present in a given medium. The instant "Bob" made contact with 'his photon' Reality was instantaneously 'recalculated' and automatically "reset" to insure that 'symmetry' remains unchanged. We've learn a hard lesson in physics and that is that **something** must be "sacrificed for the greater good."

What seems to be a contradictory assessment under the optics of present theories is precisely the answer to this mystery: the waveguide of the photon traveling to "Alice" was shifted from one harmonic to a different one instantaneously. Entangled particles are not independently legitimate. They both share a common symmetry and can't exist apart from each other unless the remaining one goes through an instantaneous 'relocation' inside a different harmonic present inside an "invisible" **waveguide** and in which a particular Reality is true for the entire system.

*****

The Moon's contribution to the non-linear nature of our Reality here on Earth added a "wild card" to our notion of physics plaguing it with uncertainties, probability waves and continuous renormalizations of mathematical infinites. Life came with a high price to pay.

Without the Moon DNA wouldn't find a non-linear Reality and complex protein chains would have been impossible to conceive.

Without the Moon the physics behind a much more simple Reality could have been explained with the theory of a simple harmonic oscillator and an easy application of a Euclidean Space Analysis but it would have been postulated by Aliens from another galaxy and not for us.

Without the moon the living world hadn't been granted with the right to exist in the first place. There couldn't be ocean tides and females wouldn't have a define cycle for reproductive purposes either.

Without the moon love would have been much less romantic and Beethoven would have never composed such a magnificent piece of classical music. Although I have a gut feeling that science don't really care much about the later two anyway.

CHAPTER II

"Post hoc ergo propter hoc"

I took the liberty to transcribe the meaning of the "pattern" symbolized by the Latin expression that borrows a title to this chapter.

The form of the post hoc fallacy can be expressed as follows:

A occurred, then **B** occurred

Therefore, **A** caused **B**

I agree that logic could be sometimes cheated by a good fallacy in the case where an argumentation was based on purely coincidental events related to each other only by their chronological order or sequence in which both events materialized. The opposite also holds true which means that we could be unintentionally blinded by the appearances of a fallacy in the case where two or more assumed coincidental events were connected in such a way that defies logic in the sense we know it.

That was the dilemma that hunted me for years until one "lucky day" (Read wishful thinking) the fallacy became logical and the logic turned to be nothing more than another fallacy.

This is the rationale behind the **experiment** that will hopefully revolutionize the way we see the physical world; the interactions between the objective **Reality** and the elements of existence.

One day I was stroke by an incredible thought. That day I found my laboratory. I had at my disposal an entire team of technicians working for me and the most sophisticated instruments to prove my theory: A Lottery game called **"Cash 3"**.

In most States of the Union Lottery is classified a legal game where day by day millions of people put their faith and prayers on the final drawing of one single ticket representing thousand and sometimes millions of possible **ODDS** against just **ONE**.

Then I found one that was perfect to the Test: **Cash 3**.

Why **Cash 3**? The answer is simple. Each 'machine' was assigned with a specific order. Allow me to expand in this particular point before passing to more complicated analysis later on.

Let's assume that the lucky number drawn this afternoon was **"357"**. It would mean that the first machine from left to right provided the result corresponding with the number **"3"** given by the 'seemingly coincidental' releasing of a plastic ball previously numbered with a **"3"**. The same way **"5"** belonged to the machine located at the center of the TV Studio and finally **"7"** the last ball released by the machine to the extreme right.

This is extremely important for one reason: Scientists can't "tag" atoms inside a crystal or molecules inside an ocean tidal wave to see their behavior with respect to each other; however I succeeded in the search of such a system that although seemingly random could provide me with invaluable insights as to whether such "invisible force" actually existed or it was just a figment of my imagination.

Until today I can't say if it all was the result of a personal intuition or a gift from the Almighty but whatever it was, it definitely changed my perceptions about the universe as a whole.

First let me take you step by step so you could see my vision as it materialized in front of my eyes. As an example, the numbers below are some of the results of the **Cash 3** game corresponding to the State of Florida within the period of April-May 2010.

04/11/10 MID 406	04/13/10 MID 228
04/11/10 EVE 642	04/13/10 EVE 342
04/12/10 MID 367	04/14/10 MID 827
04/12/10 EVE 901	04/14/10 EVE 315

As you can see I'm showing you four entire days of results including both the drawings done on midday and those ones in the evening as well. Nothing much to see really... All are random numbers with no correlative sequence whatsoever. This is the "ugly truth" we are exposed every day; no chance for a 'hint' and no signs of a 'handle' where one could build upon a winning strategy... Nothing!

This was my first failed attempt to uncover some sort of badly needed miracle pattern.

Then the miracle happened!

As I was putting all the pieces of Reality together I thought that perhaps what I was looking at was not the result of randomly chosen balls but the orderly selection of spatial points in a three dimensional space: Balls corresponding to harmonics inside a wave (!)

This was the first image that changed the whole game into a laboratory at my personal disposal. Then the second and most needed brake: The transformation of those results into **Real numbers** and **Imaginary numbers**.

I recalled that Max Plank relied on a "mathematical trick" (As he called it) to come up with the right solution while working on the black body radiation dilemma; the known Plank's constant.

Those familiarized with the classical representation of geometrical objects in 3D would recall that in order to plot one point in 3D space you need spatial coordinates corresponding to three axes; the horizontal "X", the vertical dimension "Y" and the one that gives the third dimension a background perspective to objects "Z"; their designation also varies according to the user so that's not a definitive factor either.

Let's go back to our results for the days of April and see how they look after being transformed.

406 will look like **(4,0,-1)** How did I arrived to such conclusion? First of all let me show you the mathematical way to get there.

0 and **5** both correspond to the point of origin in the coordinate system whether you use the Cartesian method or the Euclidean coordinate system in the so-called Euclidean space.

The numbers **1,2,3** and **4** are Real numbers. Where the numbers **6,7,8** and **9** arc thcir corresponding images.

This way we have **6 = -1, 7 = -2, 8 = -3** and **9 = -4.** Let's transform a couple of more so you practice the rule a little more. The following result was **642.**

Then if **6 = -1, 4 = 4** and **2 = 2** the final transformation may look like **(-1,4,2).**

When I said that 'may look like' I'm just saying that in future numerical analyses involving two or more congruent results" I'll be as free to use both the Real part and the Image part as Reality has been entitle to do ever since the first moment of Creation.

In the absence of a sophisticated software program capable of representing a given set of points in 3D space and the mathematical calculations to predict the exact answer I had to rely on purely subjective and simple tricks that so far seem to work in 100% of the cases analyzed.

Let's talk about the possible solution for a point in a 3D space system of coordinates.

The simplest case is when the point is located right on top of the origin **(0,0,0)** or **(5,5,5).**

For one point represented by the coordinates **(1,1,1)** The eight possible location inside the eight (8) sub-dimensions of our 3D world are the ones seen in the next page.

1) **(1,1,1)**
2) **(-1,1,1) or (6,1,1)**
3) **(1,-1,1) or (1,6,1)**
4) **(1,1,-1) or (1,1,6)**
5) **(-1,-1,1) or (6,6,1)**
6) **(1,-1,-1) or (1,6,6)**
7) **(-1,1,-1) or (6,1,6)**
8) **(-1,-1,-1) or (6,6,6)**

If you place the three axes (coordinate system) inside a cube, you'll find that each half of the cube will be able to contain four symmetrical points on space with their own images.

The simplest "Polygon" in analytic Geometry is a Pyramid. The symbolism implied in its shape goes beyond imagination. The Pyramid is an orthogonal polygon and the simplest object (first of a long list) representing four points in a 3D space. In other words, pyramids are the 'signature' of Reality in space. The concept engraved in the spatial properties of a pyramid synthesizes the heart and soul of Occam's razor; its message and its intentions.

Now you know the secret. You are free to apply this mathematical correspondence to the latest results published in your state of residence and become an active witness of the truly **HOLYSTIC NATURE** of **REALITY.** But perhaps even more than that... You could make some "bucks" on the side too ☺

Refer to the chapter "The Proof" to understand the 'Reality-given-connection" of the numbers and images that I mentioned to you above; namely **0-5, 1-6, 2-7, 3-8, 4-9**, they will be the key to interpret every single element in the *"Periodic Table of the Elements"* accordingly to new concepts of Reality postulated in this book for the first time in human history.

I'd like to conclude this chapter repeating once again a quote from by Lawrence M. Krauss from his book "Fear of Physics" and the same exact one I published on page 119, 4th Paragraph of my book "Warning Scientific Discretion Advised".

"...If at the scale where "Quantum Gravity" becomes sick, a brand-new kind of physical theory emerges, based in a new kind of mathematics that pushes the limits of our existing knowledge. It has been argued that yet new symmetries might become manifest causing the scale dependence of physical theory to stop. If this is the case, then the theory defined at this point can be truly called 'complete.'

CHAPTER III

"The patterns"

This chapter will put you minds at ease in regards to the question whether those patterns mentioned before are real or just a figment of my imagination. The series below are real results corresponding to the period of April-May of 2010 according to publically announced winning numbers for the **Cash 3** game. See for yourselves:

905..........406..........901..........406.......090......951.....595

As you may have already realized after some easy conversions the relationship couldn't be more obvious, so let's do it:

400..........401.....401......401......040........401.....040

If I'm not delusional I could say that I'm seeing <u>seven</u> 'links' in a chain with a very small and recurrent difference; can you not? So! I decided to assign a name to that upsetting intrusion and I chose a very well known mathematical term called **"perturbation."**

Yeah, right! Some of you may begin to believe that the only 'perturbation' in all this is not a mathematical one but a psychological one instead... Do you think I'm crazy?

To be completely honest with you the word has cross my mind more than once.

All I'm asking from you is for a "leap of faith". Give me more time and you'll see what I saw.

Remember that the digits **0** and **5** are practically the same. The image of zero is itself.

For aesthetic conveniences only I decided to place ten (10) points between each number. In most cases the number of drawings separating them varies and as you'll see later on it has a logical explanation as well.

An important detail that I want you to pay close attention to is that as in the case of this particular series of congruent numbers **905** (the first) is "short" in the value of **1; (-1)** to be more precise.

All right! Let's see another:

365.....315......013......310.....816.....536.....513

Let's do the same we did with our first pattern.

310......310......013......310......311......031......013

This same non-linearity is observable this time too but in this particular case is found in the **fifth link** of the **chain**. The **"perturbation"** was equivalent to **(+1).**

One more pattern to go and I go right away to explain what's going on inside those machines.

128......867......367.....735...317....780....713

After performing my transformation on those series the result looks as follows:

123.....312.....312....230....312....230....213

Here we find non-linearity in two chains instead of just one as it was the case in the last two examples. Those chains are the **fourth** and the **sixth** links of the chain. In both cases the **"perturbation"** was cqual to **(-1).**

By now you may start asking yourselves "What in the world I decided for the use of such a weird word as Perturbation?"

If you are asking yourselves this question I assure you my dear reader that you are neither a mathematician nor a physicist to begin with. However! I am neither of those either so I'm going to explain in simple terms why the need for those concepts.

Long time ago scientists began to realize that most events in nature respond to an inherent tendency to periodically deviate from a linear approach. Scientists and mathematicians began to adapt their theorems and postulates to fit this increasingly upsetting phenomenon which is commonly known by the concept called non-linearity. In the field of mathematics this work was called "Perturbation Theory" and it deals with a kind of renormalization or readjustment based on the addition of a constant to fit theory with results. The same story was observed in physics especially in the field of Quantum Mechanics after a number of theoretical results manifested themselves incongruent with respect to those obtained via experimentation.

But let's rely on the experts to clarify this term in a formal way. The following is a literal quote obtained from the Internet Encyclopedia:

"Perturbation theory comprises mathematical methods that are used to find an approximate solution to a problem which cannot be solved exactly, by starting from the exact solution of a related problem. Perturbation theory is applicable if the problem at hand can be formulated by adding a "small" term to the mathematical description of the exactly solvable problem. Perturbation theory leads to an expression for the desired solution in terms of a power series in some "small" parameter that quantifies the deviation from the exactly solvable problem. The leading term in this power series is the solution of the exactly solvable problem, while further terms

describe the deviation in the solution, due to the deviation from the initial problem." (End of quote).

I think is time to keep my word about giving you the reasons for the small deviation seen in those "power series" or "patterns". However allow me first to take advantage of the Encyclopedia once again as logical evidence before I introduce my conclusions:

Perturbation theory is closely related to methods used in numerical analysis. The earliest use of what would now be called *perturbation theory* **was to deal with the otherwise unsolvable mathematical problems of celestial mechanics Newton's solution for the orbit of the Moon, which moves noticeably different from a simple Keplerian ellipse because of the competing gravitation of the Earth and the Sun. Perturbation methods start with a simplified form of the original problem, which is** *simple enough* **to be solved exactly. In celestial mechanics, this is usually a Keplerian hayat ellipse. An ellipse is exactly correct when there are only two gravitating bodies (say, the Earth and the Moon) but not quite correct when there are three or more objects (say, the Earth, Moon, Sun, and the rest of the solar system)". (End of quote)**

This is what I tried to point out in the introduction of my theory. This paragraph explains in simple terms what lies beneath those uncertainties that have kept physics in the dark for so many decades. First I thought that without the active influence of our natural satellite (The Moon) these non-linear "headaches" wouldn't be a bothering us at all but the truth is that none of us would have been around either to testify about it. But let's go back to our increasingly 'boring' game of lottery.

A. If I didn't falsify those results that you were introduced to a moment ago and I promise you that I didn't.

B. If I didn't invent or alter in any way those chains and you could verify for yourselves the legitimacy of each and every single one of them.

Wouldn't be this a good time to apply the Occam's razor once again?

Shouldn't be the **simplest answer** to a **question** (usually) the **correct one**?

Before I entertain you with a true story about the experiment that brought us and modern physics into the 'wrong tunnel' I have a simple and straight question for you.

Pay attention to it since it's a tricky one:

The "Theory of Probabilities" (Mr. Theory) has mathematically predicted a highly unfair odd equal to **1 winner in1000 plays** for the lottery game of **CASH 3** or what's roughly the same thing as to say **1 winner every 999 losers!**

Let's suppose that you decided on using my method considering the correct treatment for the perturbation factor (an effect that I also observed in the behavior of those patterns with respect to the phases of the Moon as you'll see later on).

If in fact those patterns behave the way I showed you in three previous examples would you be willing to agree with me that your new odds had been decisively reduced to your advantage?

For the first sample the "power series" was **401** and after perturbation **400.**

The total number of possible combinations will be based on the following table:

4 and 9 for the "X", **0 and 5** for the "Y" and **1 and 6** for the "Z" axes respectively.

The total combinations are:

401, 901, 406, 906, 451, 951, 456 956 400, 900, 405, 450, 905, 950, 455 and 955 (Including a possible perturbation effects)

It certainly doesn't look like the same wide limit of probabilities offered to us by the "Law of Probabilities," does it?

Hold on! If you think that Miami is the only city that presents those strange patterns inside their result list I have news for you: wrong!

The following "power series" were calculated and resumed for you out of the list of results from **the State of Georgia (From 04/02/10 to 05/01/10):**

736......687.....317......763.....317......236.......817

The **"Hamiltonian"** or also called in physics **"the ground state"** of the chain is **"123"** and after conversion looks like this:

231......132......312......213.....312.....231.....123

Pretty nice!

Perturbation was left out in this particular "power series." Let's see another:

857.....275.......730.....258....587......037

The **H** for the system is "**023**" so the conversion looks like this:

302..........220..........230..........203..........032..........032

Same as seen in previous samples before where one link of the chain (2nd in this particular case) shows a **(-1)** factor of perturbation.

There is one final point that I want to make about my earlier claims that Reality is not about the presence of one single harmonic but the instantaneous presence of infinite sub-harmonics and overtones multiple of the "original one." It is this **truth** what makes possible the instantaneous response of Reality in the compromising solution between the two universal principles i.e. the Laws of Conservation of Symmetry and mass and energy but also perturbation.

This is a short example of a "Long Wavelength Harmonic". A numerical example in the Florida Lottery (CASH 3) is seen below:

367..... (25 drawings)....713..... (25 drawings)....173

After conversion looks:

312..........213..........123

This specific "long range power series" are separated by exact number of 25 drawings in between! Couldn't this be a typical example of an 'electron' orbiting an "S" type shell in atomic orbital?

I have undeniable mathematical proof that the previous assumption is not only the correct one but the simplest one! (Please refer ahead to chapter titled "The Proof")

I'm convinced that if we fed the right software into a computer instructing it to analyze those patterns we'll be amazed of what we'll find! I wish I could one day be able to see its 3D graphical representation; nothing could be more satisfying to me than finally being able to prove that the "ghost" hunting me all these years was as real as I am.

With this last example I have the feeling that "you got the message", so let's move on and draw the final conclusions.

At this moment you may start asking yourselves the same question:

-Is this your "best shot"? (You may ask).

-Of course not! (My answer directed to you)

This is just the beginning. But you must admit that we've gone a long way ahead from the miserable **1 to 1000** odds promised by the Lottery in comparison to this preliminary **primitive "power series"** that I had shown you so far, am I right or wrong?

Good! Be patient and stay with me the rest of this chapter and we all see the EXIT at the end of the tunnel.

Q. Why did I just catalogue my "power series" as primitive at this point??

A. Because they are. They only reduced those odds from an astronomical ratio of **1 in a 1000** to a better but yet incomplete one of **1 in 16** which is left to the 'mercy of the Theory of Probabilities" and uncertainty... Not enough!

The bad news is that by using either eight or sixteen combinations you are still relying on "Mr. Theory" for a probabilistic outcome.

I said I promise you a theory that will prove the "Theory of Probabilities" to be obsolete and I'll keep my promise.

The good news is that such a theory exists and it's called "Synchronicity". I've been familiarizing you with official terms as my story progress because I want you to be able to use the same "language" scientists employ in their day-by-day life.

Let me show you "something":

515.......562..265.....775.511...155!

This is a real power series! (Go to www.florialottery.com for references).

The number of points is the exact number of drawings in-between them.

Let's convert them down into their corresponding "Objects":

515.......512..215.....225.511...155!

After symmetry cancellation (Just like waves in space):

(515) with (155), (512) and (215). But here we have something very interesting! Look at *(225) with (511)!*

As you could see I never change 5 to 0! Although 0=5 since; 0 is always the beginning of each complete cycle and 5 is always the end of the first half-cycle and the beginning of the second half-cycle it also means that the reference changes too. Pay attention to this:

(225) If 5 is the reference then P= {-3-30} Pt=-6 and (511) p= {0-4-4} Pt= -8. Also if you decided to study those patterns in a closer detail you'll see that there is also a conservation in the number of points (actual drawings between them).

The difference is (-2) and we know that -2=7 so we could say that it's not over yet... Keep in touch!

CHAPTER IV

"The Double slit experiment: The holistic spectrum of our Reality"

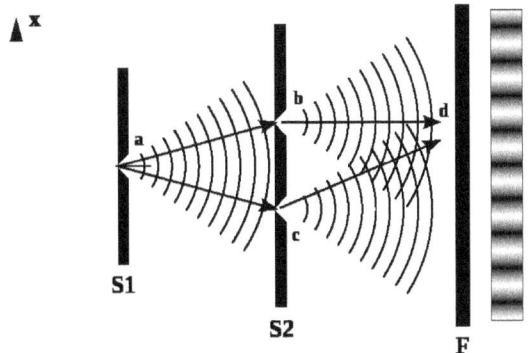

Could you see those spherical lines crossing each other out between S$_2$ and F?

Those are 'my problem'!

This was a crucial interpretation concerning an important piece of evidence becoming a theoretical disaster as the result of a stubborn adherence to the idea (Derived by the interpretations of Reality provided by Quantum Mechanics) that endows mass and energy with a quantized nature and a magic duality they never had in the first place.

Physicists are still convinced that what they observe at the screen is the result of some "virtually inexistent result of a "no-one-has-ever-seen probabilistic wave" guiding an 'electron' or a 'photon' to its "final destination." Destination being absolutely proved to be always the same no matter the amount of time elapsed between shooting of 'electrons' one at a time ☻

What's only truly quantized in this universe of ours is Reality.

Those balls of the lottery game of **Cash 3** have been exposed to the same exact "invisible forces of destiny" as once have electrons shot one by one during the famous "Double Slit Experiment?"

Those "power series" anyone may find just by following those simple steps I showed earlier still remain hidden under the list of "winning numbers" all across this country.

If there is something we all could agree on is that those repetitive patterns I've discovered hidden inside the "apparently random numbers of a game of luck" shouldn't be there at all considering the odds obtained from the "Theory of Probabilities".

The "Two Slits Experiment" (For those not familiarized with it) was originally designed by Thomas Young to prove the wavy nature of light against Newton's claims about the particleness nature of light. What it was thought to be an interference pattern between light rays on the surface of the screen became the greatest mystery among scientists and the "Achilles heel" in the field of modern physics even today.

Five years ago in a chapter titled **"How Well Do You Remember the Two-slit Experiment?"** On page 87 of by book "Warning Scientific Discretion Advised," I introduced the reader with a funny version of the subject. After giving it some thoughts I finally decided to import some of the paragraphs written back then for your personal amusement, let's see it:

"...Let's take a look to a brief comment of Paul Davies and John Gribbin appearing in a book titled The Matter Myth:

"...Each photon arrives at the image screen and makes a spot on a photographic plate. In the equivalent electron experiment, single electrons are fired through a double-slit system, and the "image screen" is a sensitive surface like that on a TV screen. The arrival of each electron makes a spot of light on the screen, and a video of the buildup of the spots of light shows how a pattern emerges as more and more electrons pass through the system."

Now it comes the tricky part!

"...Recall that one cannot know in advance, because of the inherent uncertainty of the system, precisely where any given photon or electron will end up. But the cumulative effect of many "throws of the quantum dice" will average out the distribution into a well-defined pattern..."

The dice aphorism is not mine, but actually theirs.

"...Moreover, this pattern shows the same series of interference bands as obtained with a strong source..."

Finally the puzzle!

"...Each particle, be it photon or electron, can clearly pass through one slit alone. And each particle, as the buildup of spots on the image indicates, behaves like a particle when it arrives, striking the screen in just one place. So how can an individual particle, which can pass through only one of the slits, 'know' of the existence of the other slit and adjust its behavior accordingly?"

...And now the conclusion!

"Could it be that a wave of 'something' passes through the two slits, only to collapse into a particle when its position is "measured" by the screen? This is surely too conspiratorial, for the electrons or photons would have to know our intentions. And how does each individual "know" what the others will do so it can decide where it belongs in the interference pattern that builds up?"

A self-conscience and conspirator minded particle? And we though physicists didn't have a vivid imagination...! The following words come from another physicist and author of many books on the subject, Michio Kaku from his book "Hyperspace" one of ET's favorite, by the way.

"...If quantum theory violates our common sense, it is only because nature does not seem to care much about our common sense."

I wonder why those opinions start sounding too familiar to me. Anyway, here comes ET's favorite part!

"...As alien and disturbing as these ideas may seem, they can readily be verified in the laboratory."

"...As physicist Sir James Jeans once said, "It is probably as meaningless to discuss how much room an electron takes up as it is to discuss how much room a fear, an anxiety or an uncertainty takes up." (End of quote from page 84. "Warning Scientific Discretion Advised")

So, this is it! I hope you've enjoyed this colorful introduction to a (Fair is to say) very serious and complex experiment that literally "derailed" physics in the past from its right path.

Before we proceed with the real answer as to what really happens on the screen I feel that I should clarify something said in the first quote. This is what they said in the context:

"...because of the inherent uncertainty of the system..." What they meant by this expression reflects the tricky nature of the quantum world.

There have been hundreds if not thousands of naïve technical suggestions and theoretical tricks aimed to by-pass this inherent property of the quantum world acting between us, the "Observer" and the "Observed" (Photons or electrons) with absolute **Zero** chances for success.

The moment we tried to "observe" (Tag with some sort of detectable means) which slit a photon or an electron "decided" to move through, the familiar spectrum would instantaneously disappear from the screen. "These are the facts and they are indisputable" (End of quote)

But that's just part of the problem! The tricky part is to explain how those electrons could know in advanced that they had already "reserved" their "Final Destinations".

Let's hear some versions of this issue first.

Quantum Mechanics bases its interpretation in the so-called "Copenhagen Interpretation" saying that the electron becomes real the moment we measure it on the screen via collapse of the "wave function."

What is a "wave function" you may ask. It is a virtual (Physically non-existent) wave of probabilities that is capable of performing instantaneous calculations so by the time the electron hits the screen, it already "know" where it belongs. The man who proposed this particular version (trying to represent reality) was Max Born and he was awarded with a Nobel Prize ☻

If you take a closer look at the previous quote I posted here for you, there is an interesting part I'd like to discuss with you very quickly. Remember the part mentioning the "throws of the quantum dice"?

Mr. Theory (The Theory of Probabilities) was based entirely on the postulation of this very principle.

The **Cash 3** game you try to win, the millionaire who won the Lottery and the probabilities used in the Stock Market statistics they all **bet** their chances in an *incomplete and two dimensional way of representing our 3D reality.*

Another attempt to explain this experiment and one that deserves to be mentioned here is the "Pilot Wave Theory" also known by many aliases: "De Broglie-Bohm Theory", "Bohmian Mechanics" and "The Casual Interpretation" among many others.

Let's see how the Internet Encyclopedia characterized Bohm's interpretations of what actually happens in the Two Slit Experiment:

"In de Broglie–Bohm theory, the wavefunction travels through both slits, but each particle has a well-defined trajectory and passes through exactly one of the slits. The final position of the particle on the detector screen and the slit through which the particle passes by is determined by the initial position of the particle. Such initial position is not controllable by the experimenter, so there is an appearance of randomness in the pattern of detection. The wave function interferes with itself and guides the particles in such a way

that the particles avoid the regions in which the interference is destructive and are attracted to the regions in which the interference is constructive, giving rise to the interference pattern on the detector screen." (End of the Quote)

Bohm introduced a "non-local approach" to the image seen on the screen which is in my views a step forward in the objective representation of Reality. However he felt short in the most important aspect required by any theory that dares to call itself 'the final one' and that is **predictability**.

Bohm spoke about a "local wave" or guiding wave that interacted with some sort of "non-local one" that he believed was supplied by the entire universe. The collapse of the first didn't affect the integrity of the second, according to his interpretations.

But let's see what the Encyclopedia had to say about it:

"...The basis for agreement with standard quantum mechanics is that the particles are distributed according to $| \psi |^2$. This is a statement of observer ignorance, but it can be proven that for a universe governed by this theory, this will typically be the case. There is apparent collapse of the wave function governing subsystems of the universe, but there is no collapse of the universal wave function..."

One more quote and we're done with it:

"...De Broglie–Bohm theory gives the same results as quantum mechanics. It treats the wavefunction as a fundamental object in the theory as the wavefunction describes how the particles move. This means that no experiment can distinguish between the two theories..." (End of Quotes)

As some people would say... **"-Nice try though!"**
The first quote implies a condition that it shouldn't be dismissed. This is what you read a moment ago:

"...But it can be proven that <u>for a universe governed by this theory,</u> this will typically be the case..."

The original concept of the existence of a 'pilot wave' was shared with optimism by Einstein himself when De Broglie first proposed his ideas. For Einstein even a "local wave" with some deterministic nature was desirable rather than the alternative to accept a virtual existence of a non-physical wave permeated with uncertainty while promising nothing more than probabilities.
Unfortunately for him and his theory Bohm overly exaggerated the influence exert from the entire universe in the determination and "final destination" of our electron or photon at the screen. If it's true that our Reality is in a constant state of change it's also

true that humanity has reached a high level of knowledge and understanding of the universe in general. I can assure you that the mathematics and the status of our technology is enough to formulate a theory capable of representing Reality in a time-progressive fashion.

The core of my theory and the point that I've been trying to make since my first sentence is that our universe is not governed by that theory (i.e. Quantum Mechanics). The universe does not answer to virtual waves of probabilities represented by "the square of the wave-function."

This was the dilemma started in the first Solvay Conference between Einstein and Bohr over the need to find ways to represent an objective Reality. The "old man" was practically muted by the chorus integrated by the Copenhagen Trio: Bohr, Heisenberg and the rest of the team.

The day Quantum Mechanics won that battle was the day Occam's razor was unfairly exiled from our science.

The truth is that there was no "guiding wave" and no "universal wave" either. To understand the mystery behind the Two Slit Experiment you must open your mind to a new way of thinking:

The moment energy (photon or electron) is scattered into space from a "local Reality" (i.e. Micro-world) ruling inside the atom, it will instantaneously adapt into our Reality spreading energy/mass along a different kind of wave no one has ever been able to give proofs of its existence until now:

The harmonic wave configuration of (Its) Reality present at that precise instant in the vacuum ☻

It's true that there are almost infinite number of sub-harmonics and overtones corresponding to the multiples of the elemental wavelength of Reality at that point in space and at that instant of time, but the quantum will be instantaneously "placed" into its "carrier" by a "high order of things" that can't be cheated or changed unless we intervene in the system. In other words, the so-called spectrum was already **on place** in a direct correspondence with the previously set distance of the source from the slits and the screen. Even before we shot the first electron or the first photon across the two slits Reality had already ruled its "final destination" based on two universal principles: The conservation of symmetry and the conservation of mass and energy. Energy collapsed into a point-particle with defines matter properties abandoning its wavy nature the moment the screen was hit by it.

Let's assume that you know nothing about sculpture yet you pick up a handful of plaster and throw it up in the air. As the plaster fall by the action of gravity it begins to fill from inside out a beautiful statue. You take another handful of plaster and repeat the previous procedure this time not even looking at the statue and surprisingly you realize that the statue continues to build itself with every throwing of plaster. It is as if the statue was already there just waiting for you to fill it out if you decided to.

The issue whether 'electrons' are or not particles and can or cannot divide themselves is so naïve that doesn't even deserve to be explained. Electrons are indeed real and they do become particles with all the attributes expected from matter but this "Cinderella wish" only becomes real **SEVEN TIMES** within each cycle of Reality (4 times for the "object" and equal number of times for their corresponding images) in between "They" cease to exist as matter only to become both wave of energy and partially formed matter.

One could say in simple terms that neither of both in particular yet both at the same time. As I'll try to illustrate later on with numbers Reality finds **"Itself"** in a compromising situation. Human logic is based on a ten numerical language but our universe must fit into a 3D space. **Both concepts are incongruent with each other** and one must be altered in some radical way.

Space can't be changed or deformed since Reality couldn't even make sense in such scenario so the logical explanation left for me to come up with was that numbers and the Reality meant to be represented by those numbers had to go through some sort of "sacrifice" for the "greater good".

This is in fact what happens and we'll see all the evidence you need in details in chapter VII titled "The Proof." (The selection of the order VII to this important chapter was not exactly another coincidence as you will understand much later).

As for the whimsical 'electrons' with a 'mind of their own' is nothing but a myth. The moment we placed the energy source, the double slit obstacle and the screen Reality had already set the "statue" before time became time. The plaster and the way we throw it into the air are entirely irrelevant. If we want to know the exact point of impact for each 'electron' we will have to play under the rules of Reality and stop guessing in a silent complicity with a primitive and defective mathematical tool.

The speed at which Reality (Not the universe) determines the "final destination" of our particle is a universal rule "computed" instantaneously in every single event in both worlds: The Macro-world and the Micro-world respectively.

Gravity and entanglement are both the results of an instantaneous solution for the eternal confrontation faced by Reality in a constant search for a compromise between the conservation of symmetry the

conservation of mass and energy and net perturbation of the system.

As we'll see ahead in subsequent chapters:

"Our universe is far from perfect without 'a perfect solution'. This constant seeking for compromising physical processes are the paradox and the "key" one must understand before 'he' or 'she' claims the understanding of the "Theory of Everything; reality is perfect due to the 'melting pot' of two powerful fundamental concepts." (Miguel De Zayas/2010)

My next chapter will introduce you into the microscopic "Fabric of Reality" inside the atom itself. You'll be able to understand what kind of Reality was present inside atoms before they shot those electrons of the 'Two Slit Experiment" into the screen.

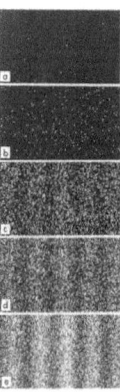

The sequence above shows the final spectrum obtained after shooting one electron at a time.

The beautiful "statue" seen at the bottom of the page was enhanced by the plaster we continuously threw against the screen; the more 'plaster' the clearer and defined the contours of the "statue."

CHAPTER V

"The real structure of Reality"

It's sad to see so many theories and opinions confronting themselves every time a new compound is synthesized in the laboratory defying fundamental rules concerning chemical bonding between atoms. I can't even begin to conceive the magnitude of the problem of having multiple systems in place trying to represent the electronic configuration of atoms and none of them being able to account for any complex situation.

Since the postulation of the Pauli Exclusion Principle until today a mystery concerning the "illogical" configuration of the mineral "Chromium" and Cooper" are among the candidates.

Those two important minerals of nature were thought to "fill in" their electron orbital following a different order than the one they did. Both are in complete denial of a theory still in used.

Let's read what they say in the encyclopedia about this issue:

"...However, chromium and copper have electron configurations [Ar] $3d^5 4s^1$ and [Ar] $3d^{10} 4s^1$ respectively, i.e. one electron has passed from the 4s-orbital to a 3d-orbital to generate a half-filled or filled subshell. In this case, the usual explanation is that "half-filled

or completely-filled subshells are particularly stable arrangements of electrons". (End of quote)

A typical "barking at the wrong tree" analogy may fit for this one too.

The absurd 'electron configuration' that as a matter of fact has nothing to do with 'electrons' in the first place but with harmonics is due to the wise intervention of Reality seeking for a compromising solution while keeping a "single harmonic pattern" able to bond with other elements. As I said earlier the so-called 'one electron configuration' is the reflection of the same pattern of a 'single harmonic' coming across a theoretical infinite number of "Standing Waves patterns" originated at the very core of the atom. As I said once before, I'll prove it to you with numbers (!)

As we saw in the analysis of **"Cash 3"**, one point in **3D space** could be represented using three axes **(X, Y and Z)** and in eight **(8)** sub-dimensions of space.

For decades scientists have been trying to figure out (Especially lately) and based of the interpretation of Reality give by Quantum Mechanics, how to arrange the atomic orbits so they could explain how they behave among each other. First they believe that each atom rearranges themselves in "shells' and "sub-shells as their number one-by-one is increased.

To understand what's truly happening one need to define Reality at that level too.

Here comes the "rules of the Game". Just as in any new theory there are concepts that we must assume "Facts" otherwise you end up with a flawed theory or what's even worse none at all.

As evidences showed in my "experiments with **Cash 3**" I'm convinced that there is an objective Reality that acts on everything regardless of scales; both in the Macro-world and in the Micro-world alike. That marvelous conclusion led me to a more remarkable one:

There are no "fixed quanta" of energy and there are no "fix mass" of point-particles either. It's all an illusion; the result of a poor judgment as a consequence of an impossible attempt to observe and study Reality in any-world. The Standard Model is anything but standard in our universe and the only thing that always determines the eventual sizes of the energy and mass involved in what we observe as "particles" is the fabric and configuration of Reality according to "Its rules".

The only "quantized thing" in this "Universe of ours" and its corresponding image (We cannot see but exists nonetheless) is Reality ☻

Borrowing my favorite author's, Victor Hugo's aphorism I could say that Reality is like a rope, made out from multitude of harmonics."

Reality is indeed indivisible and it cannot be seen until the smallest "unit" (The fundamental wavelength of the harmonic carrier of energy-mass) has completely elapsed in a **"cycle of time".**

This is precisely the reason time is not just relative but inherently unique in a given "local" configuration. In other words, the Fabric of Time is different as we go from one 'atomic shell' (As it is known) to the next just in the same way it is in the surface of Mercury with respect to the surface of Venus.

It doesn't make much more sense to speak about time or space-time as to it is to speak about electrons in different shells.

Time is nothing but a 'unit of reference'. Time becomes Time when the cycle is complete; when all the **possible** (Not probable) "moves" are "played" by "It" and all universal laws are properly fulfilled without a single violation; Time becomes Time when light becomes light in our eyes.

Time becomes Time every time you look at the mirror and see you image on the "other side."

Quantum Mechanics says that matter and energy share a "duality;" in other words mass can borrow properties attributed earlier only to energy and vice versa. This is as I said in my previous statement a poor judgment and a misinterpretation of Reality. The same mistake when we believe that a spoon could (physically) bend itself by the refraction of its image inside a glass of water.

I agree with Bohm who once said that Light was all there is out there (Assuming light in its energy definition). Matter on the other hand becomes real for the duration of eight single instants every cycle of Reality. In other words we could say that we exist (corresponding to the material definition of existence) inside an alternating and cyclical sequence for a number of eight times within the lapse of time corresponding to the whole wavelength of a quantum of light (Photon).

Let me put it even easier to understand. If we had the ability to see what happens from one instant of time to the next instant of time we would be surprised to see ourselves disappearing and appearing back into a human body just as a "turning signal" does in an automobile.

Problem is that there is a "hidden mechanism" employed by Reality, a sort of "dimensional mathematics" that I began to understand as I "traveled across amazingly simple calculations;"

and certainly one that will stunt you as it did to me when I discovered it.

An interesting experiment using sound waves, a bucket with detergent and a sound generator shows how every time resonance was achieved (simple harmonic oscillator) the exact number of eight **(8)** spheres could be actually seen forming in the surface of the water.

Those "sub-shells" referred in their theories is the configuration of waveguides accordingly with the spectrum of harmonics present in the orbit.

Imagine a waveguide of Reality just like a Rainbow. Each frequency of light separated from the next by an orderly pattern based on their frequency and energy state of the light waves.

But wait! Not a typical rainbow you're accustomed to enjoy after a tropical rain. To understand how the fabric of Reality looks like you must have a better picture in your minds...Wait! I think I found it!

If you are a man reading this book you'll probably have visited those classic Barber Shops using an especial sign for advertising; a sort of a trade mark to let you know from the distance the kind of service they provide. The "Sign" is electrical most of the time and is made out of three colors that as it rotates in a common axis gives you the false impression that is moving up or down.

Let's do together a real "Thought Experiment" like the ones Albert Einstein used to immerse himself from time to time... Here it comes:

Imagine that you could make the same Sign using other colors except for the white one. We're getting close! You also decided to assign each color with a different "radius" from the common axis. Let's say that you decided to mimic the order you saw in the rainbow and for the red (infrared) you employed the larger "radius". For the orange you used another one but with a slight shorter "radius". You continue the same procedure with the rest of the colors until you get to the dark blue (ultraviolet). At this point since you ran out of colors you decided to leave the blue in the very center of the axis with a "radius" close to zero in length.

The moment you turn on the switch you'll be the luckiest Barber in town, believe me. You may not much about hair cut but you represented the fabric of Reality in a brilliant way.

Are you able to imagine how the pattern of the image captured by the momentum of the sign may look like? Congratulations!

You'll see how the red cylinder kind of "wrapping" the others as it moves, right? Then if you pay attention to the orange cylinder you'll see that it does the same as the red did with the others except that yellow can't wrap the red.

Yellow wraps the green and the blue and green wraps the blue but when we get to the blue we see that no one is wrapping it.

We are getting closer now... Keep it with me a little longer and you'll be able to see it:

Make a circle with the "multi-color rope" and cut the rest away. You ended up with just a circle, right? Wait! I'm not done yet!

Make another rope but this time twice as wide (thick) than the previous one. Attach the ends in a circle proportional to the first and cut away the rest of the rope.

Repeat the same procedure three or four times until you get a series of concentric rings that will bear always a relationship equal to the square of the previous one.

That's the atom!

The true is that those so-called 'electron orbital's' are not the orderly inclusion of 'particles' as they fit some principles discovered by Pauli and later on stated by rules like Aufbau or Madelung, those patterns are the reflection of a Reality originated inside the core of the atom. The more patterns are formed in the atomic nucleus, the more colored cylinders will be in our "Barber Sign". As the 'fibers' inside the rope are increased by the inclusion of more internal bonding between nuclear harmonics, more complex will be the overlap between the "one-single electron" configuration and the symmetrical conservation reached by the atom with the inclusion of higher harmonics.

Now that you know the secret of how the "rope" of our Reality works let's see why this happens.

It's not a secret that the atom of Hydrogen has one 'electron' in the first orbital (1S) as they have named it using a spectroscopic approach. (Namely S, P, D, F)

The **reason** behind everything I've said in this little book of mine could be resumed in this simple fact. There is perfect symmetry in nature.

Reality builds our material existence from the bottom up and "plays around" with ways to cheat all three Conservation laws and it does it in a clever way and before time becomes time.

It is this first harmonic born in the first ring of a Standing Wave system what I call the 'fundamental harmonic of Reality'; the smallest stitches in the fabric of Reality.

Just as 'electrons' could be considered as Newton once imagined "a theoretically infinitely small point in 3D space", this uncoupled and single harmonic could explain the so-called "one-single electron configuration" for Hydrogen-like atoms..

No matter how strong and expensive laboratories we could build inside "Particle Accelerator Facilities" no amount of energy will enable us to brake such infinite force protecting this marvelous symmetry built inside atoms.

In physics there is a phenomenon known as "Tunneling" that's been observed in the behavior of certain 'electrons'. Some could even "brake through a barrier" and appear on the other side of a medium. If you apply the "Barber Sign" model to the atomic orbital you'll see how one 'electron' manages to "appear" in the other side of the barrier. The question is not if 'electrons' as point particles could do it but whether the overlap of the "an uncoupled harmonic" was large enough to be able to reach that far into the other side of the barrier. To speak about the indivisibility of an 'electron' at all times would be the same to speak about the possibilities of looking at our Sun at **03:00** Hours; it's useless, naïve and illogical. Don't misunderstand me please! It is all that I mentioned before but not because there is no Sun until it appears on the horizon but because it wasn't yet it's time to appear where it will three hours later.

The second point that I'd like to make refers to the case a proton looses energy and becomes a neutron scattering a 'beta particle' in the process. Why do you think a 'beta particle' which is nothing more than another 'electron' with a different energy signature could coexist inside a place where only protons and neutrons were thought possible?

What phenomenon is working behind "Atomic Decay"?

Oh by the way…Would we be able to know what really happened to Schrödinger's cat?

Doesn't it "ring a bell" learning that a difference between a proton and a neutron during decay was just a magnitude of energy/mass equal to the mass of a regular 'electron' besides certain properties inherent to its location inside a nucleus?

Those questions will one day find their answers after the proper understanding of our Reality.

CHAPTER VI

"The Moon, cash 3 and Perturbation"

Moons are the "Local Orbital Compensation" most planets enjoy in their own planetary orbits. The planet Mercury is to our Solar System what the single energy-mass called 'electron' is for the simplest atom of Hydrogen. It's bigger than our Moon but in some sense it is a huge Moon from the point of view that its roll is the exact same: Mass compensation as a condition of a perfect symmetry inside the core of the Sun. Just as the case in a single 'electron' around a hydrogen atom, Mercury has no Moon or "Celestial Partner" with which being able to compensate for his almost eccentric planetary orbit. It is this peculiar situation in the specific case of planet Mercury along with the explained difference in the "rope" of Reality that makes its own precession around the Sun an almost impossible event to be understood.

Our Moon on the other side has been the focal point of ancient culture like the Mayas who regarded it as a female Goddess. The earlier ancient intuition behind the roll of the Moon in the female menstrual cycle as well as the crops cycles will probably find the right answer behind the following patterns I have collected and classified for you.

Before presenting you with my "talking numbers" I'd like to explain the basic principles upon which I selected them in the first place. As I have explained earlier in a previous chapter, The Moon (just as well as **"Cash 3"**) is also affected by the same phenomenon known as non-linearity." Obviously it would have been a totally imprecise and subjective on my part if I showed you just one set of numbers (i.e. link) when it is more than obvious to assume that the effects of the Moon over a geographical zone (In this particular case Miami, Florida) could deviate from the exact time when those effects are felt much more stronger.

So I finally opted for a much better solution. I chose two "links" for the beginning of the Lunar Phase; a complementary step that could help us to pinpoint with more accuracy the effects of Reality on the **Cash 3** results, or what's the same to say the "tagged-atoms" of my 'experimental apparatus' and the Moon.

I found a great picture that will give you an idea about the 'harmonic' shape of the Moon path as it covers an entire precession cycle around our planet producing a non-linear effect in our quantum experiments and our Reality as well:

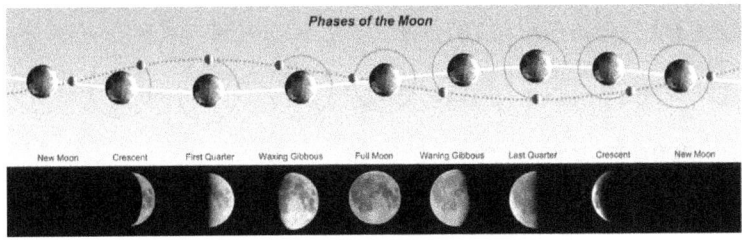

Lunar phases and Cash 3 results for that Evening:

● Full Moon ▲Waxing moon, crescent ○ new moon ▼ waning moon, crescent

Year 2009

● **05/09 (719) or (214)**

▼ **05/17 (648) or (143)**

○ **05/24 (511) or (511)**

▲ **05/31 (011) or (011)**

● **06/07 (621) or (121)**

▼ **06/15 (848) or (343)**

○ **06/22 (737) or (232)**

▲ **06/29 (999) or (444)**

● **07/07 (908) or (403)**

▼ **07/15 (908) or (403)**

○ **07/22 (260) or (210)**

▲ **07/28 (394) or (344)**

● **08/06 (714) or (214)**

▼ **08/13 (499) or (444)**

○ **08/20 (596) or (541)**

▲ **08/27 (228) or (223)**

● **09/04 (570) or (520)**

▼ **09/12 (010) or (010)**

○ **09/18 (060) or (010)**

▲ **09/26 (678) or (123)**

● **10/04 (196) or (141)**

▼ **10/11 (880) or (330)**

○ **10/18 (709) or (204)**

▲ **10/26 (374) or (324)**

● **11/02 (343) or (343)**

▼ **11/09 (113) or (113)**

○ **11/16 (638) or (133)**

▲ **11/24 (642) or (142)**

● **12/02 (364) or (314)**

▼ **12/09 (121) or (121)**

○ **12/16 (784) or (234)**

▲ **12/24 (118) or (113)**

● **12/31 (904) or (404)**

<u>Year 2010</u>

▼ **01/07 (277) or (222)**

○ **01/15 (930) or (430)**

▲ **01/23 (706) or (201)**

● **01/30 (600) or (100)**

▼ **02/05 (530) or (530)**

○ **02/14 (536) or (531)**

▲ **02/22 (498) or (443)**

● **02/28 (536) or (531)**

▼ **03/07 (820) or (320)**

○ **03/12 (820) or (320)**

▲ **03/23 (559) or (004)**

● **03/30 (906) or (401)**

▼ **04/06 (227) or (222)**

○ **04/14 (315) or (315)**

▲ **04/21 (619) or (114)**

● **04/27 (490) or (440)**

▼ **05/06 (743) or (243)**

These are the results of one whole year (from **0509/2009 to 05/06/2010**) showing an approximate effect on the **Cash 3** results under the direct effect of the consecutive phases of the moon for the time observed. I've noticed that patterns tend to show some form of recurrence and even a slight progress under an apparent influence of a non-linear order.

Let's put them in a continuous line so we could see some interesting features:

214-143-*511* or *(011)*-*011*-121-343-232-444-*403*-*403*-210-344-214-444-541-223-520-*010*-*010*-123-141-330-204-324-343-*113*-133-142-314-121-234-*113*-404-222-430-201-100-530-531-443-531-*320*-*320*-004-401-222-315-114-440-243.

Pairs of equal numbers are:

a) 511) or (011) and (011) one after the other.
b) (403) and (403) also one after the other.
c) (010) and (010) also one after the other.
d) (320) and (320) also one after the other.

The link (444) repeated after 6 exact drawings just as (113) did with the same period in between.

After two consecutive numbers we can see a third one seven plays later... example of these very common "power series" are:

a) (214)-(143) seven plays later.

b) (403) (511) or (011)-(121) seven later from the first (210).

c) (530)-(531) seven plays later not counting when (531) repeated itself (315)

Then we have the link (222) repeated itself after 13 plays; Also a typical period between equal numbers.

The same repetition we saw happening above with link (403) happens inside the atom in harmonics 6-7 as you'll see later on chapter "The Proof"

I find hard to believe that those patterns that you could find in every other **Cash 3** are just the product of a pure coincidence. On the other hand I don't see how the Laws of Probabilities could convincingly 'explain' all this to me one day anyway.

I'm convinced that these numbers and their mutual mathematical relationship will make much more sense to those directly involved in the study of non-linear mathematic sequences of numbers and perhaps will spark some interest in those involved in the field of genetics. We have learned by the studies done in biotechnology and more specifically on the DNA structure that genes are structured in codons and that those codons are selectively switched to produce enzymes, amino acids and even hormones. If my theory proves to be correct one day I bet you 10 to 1 that those effects that appeared suspicious in those patterns that I tried to present to you before have an integral part in the decision making process inside the chromosomes that make out for our DNA. It's my guess (Since I cannot prove it) that our DNA uses those cycles to synchronize their manufacturing process between different parts of the body.

It would be a basis to fundament the zipping and unzipping of DNA peptides if they had some sort of time-reference that cannot be "cheated" by any natural or non-natural intervention.

If you think that the concept of "reference" was a human invention I believe you should observe nature more often than you normally do... No offense!

Birds orient themselves by whether, animals react to nature's disaster even before they happen, plants and crops are directly connected with the phases of the Moon just to mention some examples.

In the next and last chapter of this book I'll show you the ways to make possible the dreams of an old man whose only wish was to finally find the way to represent Reality in a deterministic and exact way. I'll propose the reformulation of an old experiment and I'll prove to you that the idea of "throwing dice" would have been a thing of the past had we cherished all these years the greatest gift man could get: Logic and Common Sense.

However before I go ahead and introduce you with a plan that will allow us to predict the exact point our "unpredictable" 'electron' will hit the screen in a newly reformulated "Double Slit Experiment" setting, I owe you with a "bullet-proof scientific evidence" am I correct?

I (personally) hate numbers but if you happen to be a mathematician then my next chapter was meant for you.

CHAPTER VII

"The Proof"

Finally!

What you're going to read here will probably change the opinion you may have formed about this book regarding the inclusion in the same context of **Cash 3** with others (perhaps) more serious issues? ☻

In previous chapters I've been proposing a vision of Reality based on the configuration of the smallest "stitch" of Reality without giving you the least shred of proof. Since I cannot build a gigantic "Particle Accelerator" with the strength of the entire universe I thought to myself that ("If I put my mind to it") it had to be a simpler and cheaper way to get there.

I want you to imagine a huge clock. You know that every Earth day is measure after the elapse of 24 hours since you were just a child. Let's assume that our day is made out of two halves and each half corresponds to twelve hours each. Makes sense isn't it?

Obviously from Noon to Noon the clock must show 12 o'clock twice. I'd say that from Noon today to Midnight Reality completed half of its time and from Midnight to Noon of the next day Reality covered an entire "cycle."

Remember that although my analogy is real this is just a hypothetical example and it has been greatly exaggerated for our convenience only. I will introduce into our though experiment a new symbolism that holds true no matter how you see it.

You would agree with me that at Noon when we began our experiment the Sun was at its highest point in the sky; in Miami the heat feels worse at this time of the day especially in Summer time.

As time elapses the Sun will begin to set on the horizon until it disappears from our sight. I could assume that at Midnight the Sun followed a decreasing path with respect to the amount of light and heat generated and the relative hour of the day.

Q. Then what happens from Midnight to Noon?

A. The "Sun's cycle" we saw in one particular reference pattern will reverse. The Sun will begin its "journey" back to the same high point on the sky however instead of rising from the same place it went down (West) it did it from the opposite side (East). We could say then that the path of the Sun from Noon-to-Midnight is the exact 'image' of the path of the star from Midnight-to-Noon, agreed?

So far so good! If an Alien paid you a visit asked you about the integrity of one 'Earthy day' you could give "It" the same description as the one I just gave you.

Let's complicate things a little bit. Let's take away the numbers of the clock and change them for wave harmonics. Harmonics are made out of the simplest form of waves and they have a "code of honor" if you wish. Harmonics relate to each others in exponential ways or what's the same to say that the square root of a given number of waves is the sub-harmonic of the first.

Let's say that we could fit an exact number of waves (**4** waves to make an example) around our clock. If we decided to change our clock into the second harmonic we would have to introduce **16** (sixteen waves) of the same size of the previous ones. The simple mathematical relation looks this way: **4x4=16.** (Four times four is sixteen) which is the same to say that square root of **16** will be **4.**

So far let's resume what we have:

1) One day (A quantum of Reality) can't be divided in pieces since it represents the complete elapse of one cycle. In fact in our hypothesis it cannot be observed until the entire cycle has finished completely.

2) We saw that one cycle could be sub-divided into two equal halves being one the exact *Image* of the first.

3) We successfully replaced numbers representing hours and minutes with harmonics.

4) Harmonics behave in the exact same way as we saw with the day cycle.

If we made a comparison at the wave configuration from "quarter to nine" to "three fifteen" it would have been the same to say that its own Image could have been obtained from "three fifteen" to "quarter to nine" since we would have to move in opposite direction an equal amount of time until we complete the entire cycle.

What do we have so far? Perfect symmetry!

Fortunately for all of us that's not the case in the universe. If every single atomic orbit and every single planetary orbit around the Sun were perfectly symmetrical **the entire time** neither you nor I would be talking about this in this book.

Why? You may ask. The reason is hidden in the smallest quantum of Reality and in the very nature of Reality itself.

I'll show you a bunch of mathematical calculation (made some of them with the help of my cell phone) ☻and you'll understand the connection between **the Moon**, **Cash 3** and the absurdity embedded in every single physical theory supporting the existence of "Reality-independent-particles".

This is not a riddle. Physicists claim that if it is a fact that matter ('electrons', 'protons', etc.) sometimes behave like waves of energy it is a universal true that the very existence of particles obey to a quantum mechanics interpretation that tells us that an 'electron' ('lepton') for example is the smallest quantum of energy.

I'll demonstrate right here in this chapter those elements in nature (Hydrogen, Lithium, Iron and others) are the sole and direct product of a marvelous inter-connection between the "Cycles of Reality" and the time-progressive "materialization" of those Harmonics in each one of the eight **(8) sub-dimensions in 3D Space.**

Furthermore I'll supply you with specific coordinates of the **exact position** of those "materialization-points" identified as 'electrons' orbiting an atom (Limited exclusively to the frames of our existence) Yes, you read well!

If something Quantum Mechanics had proven to be useless it's always been its own inability to Pinpointing an 'electron' (point-mass of energy) in an atomic orbit or predicting where it will eventually hit on the screen of the **"Double Slit Experiment."**

The problem has many faces:

A) Our mathematics is in nature incomplete making out for the erroneous and probabilistic results obtained in every single field of science.

B) Our uncertainty when it comes to determine the final number of dimensions of our Reality. ("M Theory" once counted **11** (Eleven) as far as I could remember).

C) The "Theory of Probabilities" continues to be the "guess of honor" among the pillars supporting the fundaments of our knowledge.

First there are some basic rules I need to share with you so you will be part of those mathematical analyses too.

Our mathematical language is based on a small vocabulary formed by **10 (ten)** letters we call numbers.

Those are **0,1,2,3,4,5,6,7,8,9.**

Let's define those numbers according to the concept of **"object"** and **"image"** first.

Object: 0,1,2,3,4 and image: 5,6,7,8,9. Good!

The first thing you must understand is that Reality is **cyclical.** It's not a never-ending linearity as the way we count thing in our everyday life.

What I'm trying to say could be better understand if I show you how Reality elapses using those numbers or 'letters' above. This is the way Reality looks like:

"0-1-2-3-4-5-6-7-8-9-0-1-2-3-4-5-6-7-8-9-0..."

The same would be to say that half-cycle begins from **1 to 4** ending at **5** which becomes the zero of the 'image-world'; only half-way ahead. Just as **Noon** was **our zero** in previous analogies, Midnight would be **5**. From Midnight we begin our way "on the other side" called Image. We go from **6** to **9** (four steps) until we end up where we started **("0").**

Now we have a better idea of how Reality plays its rules. But this is unfortunately not the entire story.

So far we have covered the nature of what a cycle means. Now I will explain to you a different concept that is inherent to our 'material existence'.

When we count oranges we use numbers in a single "face value." There are no 'orange images' to represent. Let's put an example:

Nine complete oranges are the product of **6 + 3, 5 + 4** and **8 + 1.** Reality uses a more complex representation because **"It"** must work with both concepts, namely **"object"** and **"image"** as well.

For the sake of a better understanding we must look at the relationship found in the inherent meaning of a number.

Let's go back to the analogy of the two-halves for the 'letters' of the mathematical alphabet.

If we added **1 + 2 + 3 + 4,** the result would logically be = **10**. But what would happen on the other half (Image side) if we performed the same operation using the arithmetic concepts we normally apply when counting oranges?

Let's see it:

6 + 7 + 8 + 9 = 30. (?) Hold on!

Something is not right with this picture!

This was one thing that an ancient philosopher named Aristotle made him to believe that those last numbers shouldn't be a part of mathematics.

How could the perfect image of the first half could represent a different magnitude altogether? It simply doesn't!

The answer couldn't be simpler to understand. Those image-numbers corresponding to **6, 7, 8,** and **9** are in fact a projection of the same magnitude or quantity in a 3D space.

From the point of view of Reality 6= -1, 7=-2, 8= -3 and 9= -4.

Why? Because 5 is the same as 0 for them! Because 5+1 = 6, 5+2 = 7, 5+3 = 8 and 5+4 = 9! That's why! Pure logic!

If we were to consider two oranges as **(2.00)** it would mean that a magnitude equivalent to two has been represented in one "Axis" not even considering the other two "Axes."

That would erroneously imply some sort of deterministic "linearity sense" to mathematics and to the Reality that those very numbers were 'in charge' to represent.

When you go to the grocery store you ask for oranges and apples but unless you were suffering from some sort of mentally impairment of some kind you wouldn't dare to ask an employee for the 'image' of an orange or an apple, would you?

I know it may sound weird and perhaps even scaring to you but this is the way it is and I can offer no other choice but to accept it as the way it is and has always been.

Before I illustrate the mechanics of the "trick" employed by Reality since the first moment of Creation of the Universe I'd like to resume what we've seen so far:

a) **"0" (Zero) is the Start of the "object's cycle" and the end of the "Image's cycle".**

b) **"5" (Five) is the Start of the "Image's cycle" and the ends of the "Object's cycle."**

c) **"1" (One) is to the "Object's cycle" what "6" (Six) is to the "Image's cycle."**

d) **"2" (Two) is to the "Object's cycle" what "7" (Seven) is to the "Image's cycle."**

e) **"3" (Three) is to the "Object's cycle" what "8" (Eight) is to the "Image's cycle,"**

f) **"4" (Four) is to the "Object's cycle" what "9" (Nine) is to the "Image's cycle."**

This seemingly absurd relationship in our "linear world" makes perfect sense inside a complex fabric of calculations materialized in an instantaneous fashion allowing for the **existence** of everything we take for granted in the universe; and all that is done by the time matter and energy present their "face" to us inside a frame defined by Reality.

What I'm trying to say is that what we observe is the final result of infinite number of "mini-processes" performed by Reality under the strict rules and limitations of a long list of Conservational Variables. Chief among them are "The law of Conservation of Energy and Mass", "The law of Conservation of Symmetry" and "The law of Conservation of Perturbation." This last one is my small contribution to the field of mathematics and it will be proved in mathematical ways with the help of simple numerical examples.

Allow me to explain what those numbers below represent. I took the square root of **2** (Two) (For no particular reasons) and calculated the product of that magnitude in a linear consecutive order.

Then I applied a new method to interpret those magnitudes in a 3D context.

You'll see other terms and strange-looking calculations but my advice is... Don't be scare! It's just a 'piece of cake'

The patterns you'll see next were the simple product of the multiplication of the square root of **2** by consecutive numbers **(i.e. 1, 2, 3, 4, 5, 6, 7, 8, 9 and 10).**

Pay close attention to what happens in (8) and (9):

1. **01.4140 or (014) (140) P= {+3-30} Pt = (0)**
2. **02.8230 or (028) (230) P= {+1-10} Pt = (0)**
3. **04.2420 or (042) (420) P= {-2+20} Pt = (0)**
4. **05.6560 or (056) (560) P= {+1-10} Pt = (0)**
5. **07.0700 or (070) (700) P= {-2+20} Pt = (0)**
6. **08.4840 or (084) (840) P= {+1-10} Pt = (0)**
7. **09.8980 or (098) (980) P= {-1+10} Pt = (0)**
8. **11.3120 Or (113) (120)P= {+2-1+1} Pt = (+2)**
9. **12.7260 or (127) (260) P= {0+1+1} Pt = (+2)**
10. **014.140 or (014) (140) P= {+3-30} Pt = (0)**

Easier to see this way:

(113) (120) $|(1+1+3) + (1+2+0)| = |(5) + (3)| = 8\ |(5) < (3)| = (+2)$

(127) (260) $|(1+2+7) + (2+6+0)| = |(10) + (8) = 18\ |(10) < (8)|=$
(+2)

The most important issue treated in this book and what I believe
it's a break-through is what you see happening in Harmonics 8-to-
9. Isn't it interesting that in both cases we found (2) two "holes" in
3D space? Couldn't this 'solution' the way Reality manage to
arrange 'mass' ('electrons') in atomic orbitals? Couldn't it?

The most important issue treated in this book and what I believe
it's a break-through is what you see happening in Harmonics **8-to-
9.**

Yes! The same Harmonic pattern cease to repeat itself which
means that I found the reasons to believe that not only were both
Perturbation and non-linearity an inherent properties of our
Reality, but that no matter how you play with numbers you'll get
only **(8)** ways to represent a set of coordinates in space. **(!)**
*Notice that **10.** Give us the same results that we had when we
started; in other words **(1)** and **(10)** are one and the same point
making possible the cyclical nature of our Reality.

The physical interpretation is undeniable:

"Any set of numbers, any function operated on numbers, any
exponential or any other kind of mathematical operation

perform with them will always allow for up to seven different sets of spatial coordinates; (8) possible ways to represent a point object-image in a 3D space".(M. De Zayas/2010)

The trick was obvious: Reality managed to 'overlap' and 'recombine' different magnitudes introducing what's call perturbation in both fields i.e. Mathematics and Physics. Using the 'time' allowed by its own instantaneous nature, Reality 'worked with perturbation' to integrate both conservation Laws (Energy-Mass, symmetry and perturbation) within the frames of a 3D space.

As you'll shortly see from my next numerical demonstration Reality keeps "its promise" to represent itself in a 3D Space using seven different point coordinates which in a cyclical nature of Reality make for eight **(8)** points inside a Cube.

The Total Perturbation of the entire numerical function is **(0)** at the end. Reality is even more jealous that what we previously thought about "It". The reason I say this is because besides keeping track on energy-mass conservation and symmetry "It" also makes sure that all the changes made to create this marvelous universe of ours will be accounted for and the total sum of those "aberrations" or perturbations will **total zero (0)**. ☻

There is one small thing I must show you before going into real numbers. How could we keep on track the changes made by

Reality so we could be sure ourselves that **"It"** also kept its promise to "keep everything" in a perfect state of conservation under the principle of no rules being violated and "no bills left unpaid."

I called it the **"Mirror."** I used the same analogy many times five years ago today in my first book but I think that I never really understood the true meaning of it... until now.

So, let's see what a "Mirror" is.

I'll use parenthesis to identify each set of coordinates (The same links of a power series seen previously) only that I'm not going to represent **CASH 3** winning numbers but real numerical functions.

This will take care of how I use it but to explain when and why will take another minute so keep with me a little longer.

Thought Experiment:

Imagine that one day you look at the mirror and you found another person instead. I'm not joking! That could easily happen if we were able to divide a quantum of Reality in many fractions before TIME became TIME.

Isn't it an interesting analogy or what? Well, brace for impact because that's exactly what Reality does all the time. If my theory is correct and I believe it is, every single element in nature (Hydrogen, Potassium, Gold and Mercury to mentions just a few) are the different possible combinations of the same "pattern of behavior" for all atoms and the same model for the entire universe; no exception allowed!

In other words, the same strategy keeps playing all over again in different circumstances so by the 'end of our Earthy Day' (Within the time allowed by a quantum of Reality) we always found the same atom with the same structure as if nothing had ever happened in between.

This explanation takes cares of WHY we must keep "notes" of those small but smart changes made by Reality so we could understand how **"It"** manages to do it.

The "HOW" will be much easier. We know by now that we can represent spatial locations in 3D space by the use of coordinates and those coordinates are referred to three axes (**X, Y,** and **Z).**

Every time that we found a small change we will make a notation and we will use another parenthesis but this time we will also include a sign + or – accordingly.

We are going to use **"0"** as reference point to 'spot' the changes:

1. A conventional positive **(+)** sign for the value of 'relative increments' as we analyze it in a "counterclockwise direction" or from right to left.

2. A conventional negative **(-)** sign for 'relative increments' in a clockwise direction.

Let's see the whole thing with an example. Let's assume that you grew five **(5)** times your size from size **2** to size **32.** **{2x2x2x2x2=32}**

The number **32** will be represented in my "Mirror Analysis" in the following sets of equations:

32= (003) (200) P= {+100} Pt= (1)

As you can see I placed the "Mirror" between 3 and 2 because as you'll see often it should be placed between equal numbers, 'object-mirror' pairs and the closest we could get from any of both rules. One important note: Decimal signs and using other than the ones recommended by these two rules won't change the final result of the "power series" ☻

Let's try to explain what we see:

Obviously **3** is not "your initial image" when you were size **2**, so Reality is playing tricks with us for what we should feel fortunately otherwise we wouldn't be talking about all this in this book.

We already learned that the image of **2 was 7.** For that reason I put a **+1** in the extreme left position inside bracers. Why in the left and not in the middle or extreme right? Because we will begin always with the changes observed from <u>the closest position to the mirror</u> and from <u>inside out.</u> The zeros mean that there is no difference between the other two axes.

Another example a little more complex:

524288= (524) (288) P= {+2+4-3} Pt= (+3)

Let's begin with the "object" and "image" closer to the "Mirror" first. Those are **(4)** and **(2),** right? Then judging from right-to-left (Counterclockwise direction) we can see that **(2)** went up to **(4)** which makes an increment of **(+2).** The second (Middle distance) are **(8)** and **(2),** instead of choosing **(-6)** we will choose **(+4)** instead as the closest 'distance' under the relative value of **5.** From **(8) to (5)** becomes **(-3)** and the same we could say from **(3) to (0)** which becomes **(-3).**

So the total perturbation for number **524288 is Pt= (+3).**

Why all the notations and parenthesis with numbers appearing to us as if they already were in the simplest possible form of representation?

Isn't it number **2** enough to bring you the idea of a pair or "objects"?

I'll answer that question with another.

If I told you that you were moving away at the speed of light from an object a couple of light years from here, how could you tell?

If you were sitting inside a bus waiting to leave the station (Let's assume that you're hearing impaired) and another bus initially parked close to your window began to move backwards could you tell to yourself which bus was moving and in which direction with entire certainty?

That's the problem with our mathematics! We assume absolute magnitude to numbers and fractions because that's the way we "connect" with Reality. We treat numbers in an 'idealistic tow-dimensional plane' because we lack the natural ability to do it in a three dimensional context.

We need a special device that polarizes light so we could enjoy a good 3D movie. Our senses are not prepared to such dimensional complexity therefore we don't miss a bit their absence.

Although 3D video games are very popular among the younger generations 3D movie productions just until recently became popular after the latest premier of the blockbuster movie "Avatar". I have prepared a number of very interesting analyses using the "Mirror" to see how they project themselves in space. I even decided to study the "missing link" between the new meaning I found in them and secret codes and symbols from our ancient mythology including the Bible.

So let's begin with the simplest set of numbers we use in our daily life:

The hours of the day

When we look at our watch what is that we see?

We see a round sphere marked with **12** numbers starting from **1** for 1 O'clock (Clockwise direction) then **2** for **2** O'clock and finally we find **3** for **3** O'clock at an angle of **45° degrees**, following the same analysis we move another **45° degrees** and we have number **6** for 6 O'clock.

45° degrees later we find number **9** for **9** O'clock and finally with the same angle we return to the point of origin **(12)**. We went from **00:00 Hours** to **12:00 Hours** covering the first half of the day. Following the same trend of analysis for the second half of the day cycle we will obviously find the same exact order of numbers.

Let's represent those numbers in 3D space using a "Mirror":

1) 00:00 (000) (000) Pt = (0)

2) 03:00 (003) (000) P= {+300} Pt = (+3)

3) 06:00 (006) (000) P= {-100} Pt = (-4)

4) 09:00 (009) (000) P= {+400} Pt = (-1)

5) 12:00 (001) (200) P= {-100} Pt = (-1)

6) 15:00 (001) (500) P= {+100} Pt = (-4)

7) 18:00 (001) (800) P= {-200} Pt = (+3)

8) 21:00 (002) (100) P= {+100} Pt = (+1)

9) 24:00 (002) (400) P= {-200} Pt = (-2)

The total factor of perturbation for the complete system is **(-5)**. If this was true the universe wouldn't exist in the first place so what's the "solution" Reality always rely on?

Well... We must see how Reality deals with the **symmetrical side** of the issue at hand while wisely tries to match up those series we established inside parenthesis:

1) **03:00 Hours match (003) with 18:00 Hours (800)**

2) **06:00 Hours will match (006) with 12:00 (001).**

3) **09:00 Hours will match (009) with 24:00 Hours (400)**

4) 12:00 Hours will match (200) with 21:00 Hours (002) since it already matched (001) with 06:00 Hours no need for the second match.

5) 15:00 Hours will match (001) 18:00 (001).

6) 18:00 Hours had already matched (800) with 03:00 (003) on top and (001) with 15:00 Hours in the previous analysis.

7) 21:00 Hours did the same with (002) in (4) however there is one pattern it cannot match (100).

8) 24:00 had already matched (400) in (3) but it cannot match the second pattern (002).

Let's put both unmatched patterns together to see what happens:

(100) (002) P= {00-1} Pt = (+1) or [+2-1=(+1)]

What? This is better than -5 but it wasn't 0.

The solution to this paradox took me one more day to finally figure it out but now the puzzle is finally answered bringing along the answer to whether is our existence is unique or are we sharing this huge space with similar worlds only separated by an infinitely small deviation in Time...

If everything in nature has its own 'image' as we have in front of a mirror why would God be different?

Why one single 'object-universe' without its 'image'?

If one thing physics has taught us is that the universe is diverse and complex but still there is an 'intelligent order' [a design] behind all this marvelous diversity. If you ever had doubts about this perception of "a supreme order of things" ruling everything at all times take a look at my next evidence.

I will mention as an evidence to prove my case a puzzle; a dimensionless constant that has kept our greatest minds partner and an 'electrons' has also its own in the field of physics and generally speaking the entire scientific community in a theoretical impasse for the last decades.

Yes... It is about 137!

When I created my **"Mirror Analysis"** I decided to unveil the centuries-old mystery behind the true meaning of **137** assuming that in fact was any or just another 'fantasy' to "fish out" new students into the field of Physics. Then I realized that the "Mirror" was a success. It worked.

I knew that I couldn't place it between **1** and **3** because it won't work (As it was the case with **Pi=3.1417**). So I decided to place my 'Mirror' between **137** and **035.** That would be, according to my rules, consistent with Reality; I could say that I was eagerly expecting a "Sign" or "Miracle" to prove right my theory and prayed for this moment to be the one.

IT HAPPENED!

Let's see the miracle that "made my day":

137.03597...

Let's place the 'Mirror' to see how this number sees itself on it:

(137) (035) P= {-30-4} Pt = (-7)!!!!!!

My seven points! One seventh of the "Whole"; one of the seventh needed to represent one point-mass in a 3D space and one of the seven different absolute coordinates needed to represent a cleverly shrunk cycle of ten into a perfect cycle of seven.

Sun-Mon-Tue-Wed-Thu-Fri-Sat and the back to Sun!

Using one of the corners of a Pyramid as "a unique point of departure and origin" in a cycle 137 was one of those points!

Then I decided to place the three digits at the left of the decimal point in front of my "Mirror" to see how they see themselves:

(000) (137) P= {-1-3+4} Pt = (-1) -As I'll explain later on, 'irrational numbers' like most of the mathematical constants

used in physics show Pt= (-1) after performing on them my 'Mirror Analysis" (!)

The same result obtained when analyzing simple numbers against the mirror. We obtained (-1) in every single column. 137 taken without the rest of the fraction will represent the constant perturbation empirically calculated earlier; the same **perturbation constant for each universe; one with (-1) and its image with (+1).**

I knew it!

It was what I expected to be... God' promises finally fulfilled!

137 was "Its" guarantee that every single time we look at the mirror we'll see ourselves and not someone else's image. Every time we wake up in the morning the Sun will rise from the East and our homes and families will be there just as we left them the night before; the assurances that no matter what happens in between "It" will be there to watch over us today just as well as "It" did yesterday.

I apologize to the younger generation of physicists when I previously referred to the oldest generation of physicists but I cannot forget the beautiful life-long "psychological curiosity" toward the meaning of **137** impersonated by one of my greatest and most admired physicists in the world; Dr. Wolfgang Pauli.

"In 1958, Pauli was awarded the Max Plank Medal. In that same year, he fell ill with pancreatic cancer. When his last assistant, Charles Enz, visited him at the Rotkreuz hospital in Zürich, Pauli asked him: "Did you see the room number?" It was number 137. Throughout his life, Pauli had been preoccupied with the question of why the fine structure constant, a dimensionless fundamental constant, has a value nearly equal to 1/137. Pauli died in that room on 15 December 1958." (End of quote).

...And he never got the chance to find out what the true meaning of Reality was either.

However he knew there was more to understand about Reality than his work meant in the development of Quantum Mechanics. He always felt a deep connection with Reality that he was never able to translate and I'm convinced that it was that encrypted message that caused his hallucinated dreams to recur over and over again and not some psychological illness as history sometimes has a tendency to record...

I can't help myself but to react with an emotional response every time I recall the incident that took place in 1927 at the Solvay Conference:

"Werner Heisenberg [in *Physics and Beyond*, 1971] recollects a friendly conversation among young participants at the 1927 Solvay Conference, about Einstein and Planck's views on religion.

Wolfgang Pauli, Heisenberg, and Dirac took part in it. Dirac's contribution was a poignant and clear criticism of the political manipulation of religion that was much appreciated for its lucidity by Bohr, when Heisenberg reported it to him later. Among other things, Dirac said: "I cannot understand why we idle discussing religion. If we are honest - and as scientist's honesty is our precise duty - we cannot help but admit that any religion is a pack of false statements, deprived of any real foundation. The very idea of God is a product of human imagination. [...] I do not recognize any religious myth, at least because they contradict one another. [...]" Heisenberg's view was tolerant. Pauli had kept silent, after some initial remarks. But when finally he was asked for his opinion, jokingly he said: "Well, I'd say that also our friend Dirac has got a religion and the first commandment of this religion is 'God does not exist and Paul Dirac is his prophet'". Everybody burst into laughter, including Dirac. (End of quote)

What a remarkable man and what a beautiful portray of a "battle of ideas" between those titans in their own fields expressing their own views about a Reality they worked tirelessly to unveil.

To learn more about what I found hidden in **137** refer to my chapter **"The mystery of 137"**.

The next example illustrates how Reality works with those same numbers were destined to represent everything in the whole universe.

NUMBERS IN 3D SPACE:

Let's go to our everyday numbers:

a) **1 (001) (000) P= {+100} Pt=(+1)**

b) **2 (002) (000) P= {+200} Pt= (+2)**

c) **3 (003) (000) P= {+300} Pt= (+3)**

d) **4 (004) (000) P= {+400} Pt= (+4)**

e) **5 (005) (000) Pt= 0**

f) **6 (006) (000) P= {-100} Pt= (-4)**

g) **7 (007) (000) P= {-200} Pt= (-3)**

h) **8 (008) (000) P= {-300} Pt = (-2)**

i) **9 (009) (000) P= {-100} Pt = (-1)**

j) **0 (000) (000) Pt = (0)**

I believe that numbers speak for themselves in the analysis above.

Let's continue to see what happens as things get more complicated:

1. **11 (001) (100) P= {000} Pt = (0)**

2. **12 (001) (200) P= {-100} Pt= (-1)**

3. **13 (001) (300) P= {-200} Pt = (-2)**

4. **14 (001) (400) P= {-300} Pt = (-3)**

5. 15 (001) (500) P= {-400} Pt = (-4)

6. 16 (001) (600) P= {000} Pt = (0)

7. 17 (001) (700) P= {+400} Pt= (+4)

8. 18 (001) (800) P= {+300} Pt = (+3)

9. 19 (001) (900) P= {+200} Pt = (+2)

10. 20 (000) (200) P= {-200} Pt = (-2)

11. 21 (002) (100) P= {+100} Pt = (+1)

12. 22 (002) (200) P= {000} Pt = (0)

13. 23 (002) (300) P= {-100} Pt = (-1)

14. 24 (002) (400) P= {-200} Pt = (-2)

15. 25 (002) (500) P= {-300} Pt = (-3)

16. 26 (002) (600) P= {-400} Pt = (-4)

17. 27 (002) (700) P= {+500} Pt = (0)

18. 28 (002) (800) P= {+400} Pt = (+4)

19. 29 (002) (900) P= {+300} Pt = (+3)

30. 30 (003) (000) P= {+300} Pt = (+3)

31. (003) (100) P= {+200} Pt = (+2)

32. (003) (200) P= {+100} Pt= (+1)

33. 33 (003) (300) P= {000} Pt= (0)

There are many other topics I'd like to touch before this book ends. I have the feeling that for you (A simple reader) and for mathematicians alike this new analysis will open some 'closed doors' into mysteries that have been hidden from us for centuries. Especially knowing that **137** is the 33rd **prime number and a strong prime** in the sense that it represents more than just the **'arithmetic means'** of its two neighboring primes. Using two radii to divide a circle according to the golden ratio yields sectors of approximately 137° (the golden angle) and 222°.137 is a strictly non-palindromic number and a primeval number.

One final calculation and we're done.

The following is a partial "Mathematical Spatial Analysis" for number "2" as its magnitude exponentially increases:

a) **2x1=2 = (002) (000) P= {+200} Pt = (-2)**

b) **2x2=4 = (004) (000) P= {+400} Pt = (-4)**

c) **4x2=8 = (008) (003) P= {-20-3} Pt = (+5)**

d) **8x2=16 = (001) (600) P= {000} Pt = (0) *Notice that between "object" and its corresponding "image" Pt always is zero.**

e) **16x2=32 = (003) (200) P= {+100} Pt =(+1)**

f) **32X2=64 = (006) (400) P= {+200} Pt =(+2)**

g) **64X2=128 = (012) (800) P= {+4+10} Pt = (+5)**

h) **128x2=256 = (002) (560) P= {-3+40} Pt =(+1)**

i) **256x2=512 = (051) (200) P= {-1+50} Pt =(+4)**

j) **512x2=1024 = (001) (024) P= {+1-2-4} Pt =(-5)**

k) 1024x2=2048 = (204) (800) P= {-40+2} Pt =(-2)

l) 2048x2=4096 = (040) (960) P= {+1-20} Pt =(-1)

m) 4096x2=8192 = (081) (920)P= {+2-40} Pt =(-2)

n) 8192x2=16384 = (163) (840) P= {0+2+1} Pt =(+3)

o) 16384x2=32768 = (327) (680) P={+1-6+3} Pt =(-2)

p) 32768x2=65536 = (065) (536) P=0+3-6} Pt = (-3)

q) 65536x2=131072 = (131) (072) P={+1-4-1} Pt=(-4)

r) 131072x2=262144 = (062) (214) P={0+5-4} Pt = (+1)

s) 262144x2=524288 = (524) (288)P={+2+4-3} Pt= (+3)

t) 524288x2=104857 = (048) (570)P={+3-30} Pt= (0)

u) 104857x2=54975581 = (054) (975) P={-5-2+5} Pt= (-2)

v) 54975581x2=1099510 = (109) (951) P={0-50} Pt= (+5)

w) 1099510x2=604461 = (604) (461) P= {0-6+5} Pt= (-1)

x) 604461x2=1208922 = (120) (089) P= {0-6+2} Pt= (-4)

y) 1208922x2=73074620 = (730) (746) P= {+3-1+1} Pt = (+3)

z) 73074620x2=1460460 = (460) (460) P={-40+4} Pt= (0)

At this point we can say that the symmetrical function for the spatial representation of the exponential of 2 in 3D achieved symmetry at (460) (460) (!)

Now let's match one by one the patterns to verify that they match and preserve symmetry at the end. We will begin from (a) to (z):

1) (a) with (e) (002)

2) (b) with (f) (004)

3) (c) with (e) (008)

4) (c) with (g) (003)

5) (d) with (f) (001)

6) (d) with (f) (600)

7) (e) with (c) (008)

8) (e) with (a) (200)

9) (f) with (d) (006)

10) (f) with (d) (001)

11) (g) with (i) (012)

12) (g) with (k) (800)

13) (h) with (a) (002)

14) (h) with (p) (560)

15) (i) with (p) (051)

16) (i) with (h) (200)

17) (j) with (p) (001)

18) (j) with (k) (024)

19) (k) with (j) (024)

20) (k) with (g) (800)

21) (l) with (u) (400)

22) (l) with (v) (960)

23) (m) with (p) (081)

24) (m) with (u) (920)

25) (n) with (q) (163)

26) (n) with (t) (840)

27) (v) with (v) (901)

28) (z) with (z) (406)

29) (v) with (u) (054)

30) (u) with (v) (975)

31) (r) with (x) (120)

32) (r) with (y) (214)

33) (v) with (s) (975)

34) (w) With itself.

As you could see this process gets very complex without the use of a computer. The main goal is to be able to match each set of coordinates with another matching object, image or both. Once the process is completed and all corresponding sets are properly cancelled out we could say that Reality was able to fit all those variables within the time one quantum of Reality elapses entirely (One whole wavelength of its harmonic). Once the process matches on both sides (both reference "mirror sets") and the total perturbation down to **(-1)** we could say that the entire equation is in balance and perfect harmony.

The wavelength of the "Quantum of Reality in each orbit or standing wave patterns around the atom will be equal to the speed of light to that particular reference system.

There was no coincidence that I decided to use the exponential function of two. Atomic orbits respond to that configuration as you know. 'Electrons' are believed to be arranged in orbital in a number of two per sub-shell. I hope that this simple analysis could prove that we are able to calculate without a shred of doubts the exact position of a point-mass on its orbit and the status of its duality (Wave-particle) at all times. In semiconductors the so-called holes is the same transfer of energy only in a lower energy level. Remember always that we cannot speak of 'electrons' unless we have determined it exact position. The position of the 'electron' can be graphically represented by the coordinate that I've showed you above; the higher the number of orbits the greater the number of those "sets" going through cancellation. Those sets are the only possible place in the orbit where energy becomes one infinitesimally small 'point-mass.'

Elements of the Periodic Table of Elements are also grouped in columns of seven.

Take a look at the numbers of the left extreme side of the picture. Seven, right?

The meaning of the limit of seven (eight) finds the reasoning behind the analogy I gave you earlier about the cycle of the Sun and the seven **(7)** different coordinates needed to complete the cycle of Reality. Don't forget what I told you! Since a cycle always returns to the point of origin you do not need eight **(8)** different coordinates or specific time-point; each "Object-coordinates" will have its own "Image-coordinates."

<u>Tips for a graphical representation of the atomic orbitals:</u>

1. All three axes **(X, Y,** and **Z)** will have the positive side marked with the numbers **1, 2, 3,** and **4** and the negative side with **6, 7, 8,** and **9!**

2. When calculating Perturbation we must operate with the numbers according to their relative position from the "Mirror".

Example:

(321) (123) -In this simple case we operate 1 with 1 (Closest to the mirror), then 2 with 2 middle position and finally 3 with 3 farther from it.

3. The intrinsic value of the "objects" (1, 2, 3 and 4) equals the opposite sign of the same value for the "Image."

Example: 1= -6, 2= -7, 3= -8 and 4= -9.

We will see later on the demonstration of what 137 actually represents and this latest rule will help us to answer it.

CHAPTER VIII

"The theory of Certainty"

If Einstein's "General Relativity Theory" had a message it was that no observation of Reality can be done without a reference system. He proposed dozens of thought experiments delivering the same message all along. A vivid example is the case when you sit on a train that appears to be moving ahead as the train next to yours begins to move backwards in relation to your position at rest. Humans are endowed with something called "free will." Most of the time we exercise our own "will" regardless of the consequences our own acts may cause to others or to the environment. In the absence of evidence about the existence of a Reality that rules every single function of our cells and every single event in nature we look more and more like a disconnected piece of Reality trying to make sense of its existence while ignoring the ties that connect us to the rest of the universe.

How could a universe so jealous about keeping its symmetry and harmony, a universe so strict when it comes to compromise about conservation of mass and energy could be so careless to rely on a set of probabilities extracted from a pair of dice...

We found ourselves lost inside a "tunnel" a long time ago. There were some who believed that if we dropped some of the extra weight we were carrying with ourselves perhaps we could even move faster and find the exit more quickly. Some decided to get rid of the Occam's razor as first choice and they did it right after the decision was made. We are getting closer to the end of the tunnel and the exit cannot be seen anywhere near.

I think it's time to go back to the point we lost our ways and pick up our gifts left behind in a remote past.

Ladies and gentlemen allow me to propose the solution to the experiment that derailed our path and where it all began: The Two Slit Experiment reloaded.

I firmly believe that if we introduced changes to the original setting of the experiment we could predict its results with 100/% effectiveness.

We must have a three dimensional representation of Reality and the only way to achieve it is by having three equal sources in equal conditions just like the Lottery did with the three machines in the game of **Cash 3.** ☻

We must duplicate the same experimental setting as the one the Lottery uses daily in the **CASH 3** games. Results of a continuing shooting of 'electrons' at the screen will have to be fed, analyze and storage for further study in a fast computer with an appropriate software.

When an 'electron' or 'photon' hits the screen it becomes a point of mass in a Euclidean plane or system of Cartesian coordinates system. The screen could only be represented as a plane and the trajectory of the energy-mass will be corrected by Reality to compensate for the missing spatial coordinate in 3D space.

This is why we need three identical setting. Each setting will provide with two spatial coordinates in a Euclidean plane. The simplest way to represent a point-particle in a plane is by the determination of two axial points. In order to pinpoint the final position of an 'electron' in advance we need to have at least three sets of two-planes in a 3D coordinate system.

That is the reason why the prediction using a couple of dice is simply impossible.

It's my belief that this is the simplest way to represent a 3D Reality in a perfectly predictable system of reference.

As the input on the shootings begin to arrive for processing in the computer a homogeneous and predictable pattern will rise from the seemingly chaotic results. Reality will finally be isolated from its uncertainties and the old experiment will finally put to rest in a museum where it belongs.

We shot one machine at a time and compute each result as each electron hits the screen. We must determine some reliable and exact method to measure the exact point of impact on the screen.

As each cycle is properly fed in a computer the same spatial configuration and mathematical series of results will begin to take shape in a time-progressive format. As more and more bits of information build up at a given point in time the computer will be able to "predict" with an increasingly rate of exactitude the "final destination" for consecutive shots in the future. It's my guess that the negative effects derived from perturbation caused by the relatively changing position of the Earth, the Sun and the Moon will be a depreciable factor in the experiment.

However I'm convinced that a continuous experiment of this kind enable to collect, process and storage DATA during the entire cycle of the lunar month in the lapse corresponding to a complete 'Solar year' would be of tremendous value in the final determination of our equations. Computers will learn how Reality works both in the quantum world and in the Macro-world. I know there is an incredible amount of data in those **Cash 3** results obtained in various States but I also know that the knowledge hidden behind those seemingly random results will help us to understand the world in which we live and teach us about the connection that we have with nature and the entire universe.

CHAPTER IX

"The mystery of 137"

You don't have to be an expert to realize that the only thing those 22 mathematical operations share is the number **137**!

Before I explain to you how and why I was able to "extract" from **137** its 'secret' identity I'd like to answer a question that has been around forever:

Are we really alone in this universe?

We are not alone even in the confines of space you and I are occupying at this very moment (!)

I'm convinced that our universe is filled with multi-levels of existence coexisting together in the same space but separated by an incredibly small frame of time in between. Why can't we see it? The fastest speed of our observation can't beat the speed of light for the conditions of the so-called 'vacuum space'. Each "existence" is based on the same model explained on previous chapter and where all of the three conservation laws are never violated.

Remember that I proved to you that perturbation always return to a zero net value no matter what mathematical function you decided to "mirror"? The same way our laws are their laws and it is the inviolability of those universal principles that keeps us isolated in our own Reality preventing us to get a glimpse at neighboring Realities.

Is the possibility of an overlap a real one? I tend to believe so. I keep an open mind about those who believe that in some rare but well documented cases the cross-over between "beings" or images of them have been happening across time.

What you're going to see next is the mathematics that offers a logical basis to support such assumptions. In a previous chapter we found a discrepancy regarding the Net Perturbation of Time as it elapses into the 'eternal infinite'. If you are one of us who believe that everything could be explained by using the two most valuable gifts given to us by God then you had every reason to be worry. However I couldn't simply "rest my case" knowing from the bottom of my heart that not even "IT" (Reality) was allowed to violate its own rules. I know that for some people it may sounds illogical to understand that principles are precisely what they are and cannot be in any possible way violated; not even by "Our own Creator".

The following is the solution for the incongruence observed as Time "goes by."

The mathematical prove will be very simple; I'll arrange each 'existence' below the next in a consecutive order:

01234567890123456789012345678901234567890
01234567890123456789012345678901234567890
01234567890123456789012345678901234567890
01234567890123456789012345678901234567890
01234567890123456789012345678901234567890
01234567890123456789012345678901234567890
01234567890123456789012345678901234567890
01234567890123456789012345678901234567890
01234567890123456789012345678901234567890
01234567890123456789012345678901234567890

You probably understand by now the meaning of what I'm doing. The magnitudes have been represented by simple numbers but they could be anything; mathematical functions, equations or any other form of representation of our Reality.

I'm going to take six **(6)** numbers in any column and place the "mirror" in between to see how Reality manages to keep the overall Perturbation (Net Pt) in zero.

My first column will begin with **(210)** the first three (vertical) numbers from left to right.

1) (210) (987) {+1+3-5} Pt = -1

2) (321) (098) {+1+3-5} Pt = -1

3) (432) (109) {+1+3-5} Pt = -1

4) (543) (210) {+1+3-5} Pt = -1

5) (654) (321) {+1+3-5} Pt = -1

6) (765) (432) {+1+3-5} Pt = -1

7) (876) (543) {+1+3-5} Pt = -1

8) (987) (654) {+1+3-5} Pt = -1

9) (098) (876) {+1+3-5} Pt = -1

10) (109) (876) {+1+3-5} Pt = -1

However if we started from the bottom up as in the perfect image the results would look the following way:

1) (012) (345) {-1-3+5} Pt = +1

2) (123) (456) {-1-3+5} Pt = +1... And so on.

This is empirical evidence that proves that our dimensional existence cannot exist on its own without its perfect image. There has to be another existence occupying the same space as ours but in a lapse of time so small that could be taken as an instantaneous time deviation impossible to detect.

God also has an image and both are represented in the ancient "Yin and yang" symbol inside the popular Chinese spiritual context.

I finally found the answer to its mysterious origin.

"..You might say the "hand of God" wrote that number, and "we don't know how He pushed his pencil." We know what kind of a dance to do experimentally to measure this number very accurately, but we don't know what kind of dance to do on the computer to make this number come out, without putting it in secretly! -Richard P. Feynman (1985)

It took me less than a minute using my "Mirror Analysis" to finally understand the meaning of the mystery behind **137** which in terms of the instantaneous nature of our Reality you could say that it took me almost an eternity...

Let's analyze then this symbolic constant that appears in almost every single aspect of our universe; from the very small to the very big, from the 'electron' to 'galaxies' and star constellations alike.

We are going to analyze the constant using the two different faces of it. I'm going to use **0.08542455** and also **137.03597** and I'll arrive to the same results:

Let's begin with 0.08542455 as a sincere gesture of admiration that I've always had for the late Dr. Richard Feynman.

0.08542455

a) (085) (424) {+1-4-4} Pt = (-7)

b) (854) (245) {+2+1+3} Pt = (+7) *Remember that this is equal to 1/137 (!)

Now let's do the same with "its other face" too:

137.03597

a) (137) (035) {-30-4} Pt = (-7)

b) (370) (359) {-3+2-6} Pt = (-7)

Some sophisticated approximations have been the focal point of certain 'complains' about the lack of decimal fractions beyond this point, so I'm going to show them that it doesn't make a difference about God's mind ☺

137.035999084...

a) (359) (990) {0-4-3} Pt = (-7)

b) (137) (085) {-30-4} Pt = (-7)

Is it enough for you to see the need for a reference when dealing with numbers and especially those with such an important scientific value like this essential constant?

Since -7 is the inverse of **7 or 1/7,** I could only imagine that this constant is referring to one magnitude in a whole made of seven; or what's the same to say that whatever it represents in physics is just a seventh of a its net value.

However if we took just the first three numbers of the Alpha constant **(137)** and performed the "Mirror Analysis on it you'd be amazed about its result:

(000) (137) {-1-3+3} (-1)!!! [Irrational by nature!]

As you may have realized by now the famous symbolism represented in the three digits **"137"** was fatally incomplete to visualize the (-7) Perturbation that it implies. The correct way to refer to it would have been **"137085."**

Of course, before the publication of this book who would have guessed that it was going to "go through a series of analyses performed within "a looking glass" in order to understand its true meaning...?

This was the same value we reached earlier without taking into consideration the image of our own existence while moving ahead in TIME in a progressive fashion. (An impossibility of course).

Perhaps even more exciting and encouraging that having finally unveiled the mystery behind **137** was to find the connections between **137.03597** and **0.08542455** using my "Mirror Analysis."

Remember when I explained to you the relationship between "Object" and "Image" and between those number I consider to be the "Objects" and those classified as their corresponding "Images"?

Very well! Take a closer look at both numbers and tell me what you see!

03597 finds an image on **08542** isn't it correct? Furthermore, due to complexity involved in the practical aspect of getting those numbers one could assume that the last four approximations found in 0.085424**455** (Highlighted for convenience) could be taken as **5**. The same approach could be undertaken from the opposite direction assuming that the approximation that follows 7 in 137.03597... (Also highlighted for convenience) may be considered closer to **4**.

Watch this!

I'll arrange both numbers in a column as follows:

137.03597

000.08542

Now I'm going to "Mirror Analyze" both at the same time. I'm going to take (137) from the first number and 'MIRROR IT' against (085) first:

(137) (085) P= {-3+5-4} Pt = (0)

(013) (597) P= {-2+2-7} Pt = (-7)

(013) (542) P= {-2-3-2} Pt = (-7)

(000) (597) P= {-5+1+4} Pt = (0)

(000) (085) P= {0+2+5} Pt = (+7) **(Inverse value) = (-7)**

(000) (854) P= {+2-5-4} Pt = (-7)

(000) (542) P= {-5-4+2} Pt = (-7)

I think you got the idea that no matter how we combine those numbers we will obtain the same information. Codata:

$$\alpha = \frac{e^2}{\hbar c \; 4\pi\varepsilon_0} = 7.297\,352\,5376(50) \times 10^{-3} = \frac{1}{137.035\,999\,679(94)}.$$

If you take (137) from the expression **'mirroring it'** against the image of the second triplet (999) or (444) one obtain:

(137) (444) P= {+3-1-3} Pt = (-1)

Taking the next triplet **(359) 'mirroring it' (996)**:

(359) (996) P= {0-4-3} Pt = (-7)

Taking **(370)** and **'mirroring it'** against the second triplet to the right **(996)** one obtain:

(370) (441) P= {-4+3+2} Pt = (+1)

Notice that the difference between signs was due to moving our 'mirror' one step to the right from (137) to (370) (Moving in time ahead).

My next calculation is a mathematical prediction that perhaps one day not far away could be verify by more perfect CODATA measurements:

Let's put the number back on the screen first...

$$\alpha = \frac{e^2}{\hbar c \, 4\pi\varepsilon_0} = 7.297\,352\,5376(50) \times 10^{-3} = \frac{1}{137.035\,999\,679(94)}.$$

Let's go way ahead in the fraction... Let's say I'll take (999) and 'mirror it' against the image of the second [incomplete] triplet (94...) or (49...)

According to my "Mirror Analysis" The next fraction must be an 8 (!).

Look:

(999) (493) P= {+50-4} Pt = (+1)! Remember that we are dealing with an image so (+1) will become (-1).

The future CODATA corresponding to the Alpha constant of our universe must be:

137.035999679948!

The universe doesn't really care about decimal points as we do.

Reality only "works" with triplets that can be represented in a 3D space and must comply with all three laws of conservation; symmetry, mass-energy and perturbation.

We pay too much attention to fractions and decimals because we live in a 'MATERIAL WORLD' where the only "images" we deal with on a daily basis is ours alone in front of the mirror...assuming we ever get a time for it.

As I was putting these thoughts in words and typing it on my laptop I couldn't help myself and I began to laugh; just like that.

I was thinking back in time when I came across Dr. Feynman's favorite quote of mine, let's hear it:

"You know how it always is, every new idea, it takes a generation or two until it becomes obvious that there's no real problem. It has not yet become obvious to me that there's no real problem, therefore I

suspect there's no real problem, but I'm not sure there's no real problem." (End of quote).

I found his remarks so funny that I used it in my first book too; it can be found on page 19th 3rd paragraph.

Dr. Feynman was referring to a personal psychological conflict derived from his own inability to accept the model of Reality given by Quantum Mechanics and its physical interpretation concerning the acceptance of uncertainty and probability as an unavoidable state of mind.

Let's relax a little bit from those numbers and read a couple of quotes I found really interesting.

"The mystery about α is actually a double mystery. The first mystery — the origin of its numerical value $\alpha \approx 1/137$ has been recognized and discussed for decades. The second mystery — the range of its domain — is generally unrecognized". (Malcolm H. Mac Gregor, Malcolm H. Mac Gregor (2007), *The Power of Alpha*, World Scientific, p. 69)

"If alpha [the fine structure constant] were bigger than it really is, we should not be able to distinguish matter from ether [the vacuum, nothingness], and our task to disentangle the natural laws would be hopelessly difficult. The fact however that alpha has just its value 1/137 is certainly no chance but itself a law of nature. It is clear that the explanation of this number must be the central problem of natural philosophy" (Max Born, Arthur I. Miller (2009), *Deciphering the Cosmic*

Number: The Strange Friendship of Wolfgang Pauli and Carl Jung, W.W. Norton & Co., p. 253)

It's been part of numerology and mythology for centuries the notion that number seven **(7)** also represents perfection and in some theological circles also God's perfection. Is a fact that in some ancient 'Sanskrit Languages' "The Creator of the universe" was believed to be always **16** years old, symbolizing by that numbers the nature of its eternal existence (Which I agree a **100%**). In **16 (1)** is the "object" and **(6)** its "image" as I explained in the beginning of this book.

"The Sanskrit language, whatever be its antiquity, is of a wonderful structure; more perfect than the Greek, more copious than the Latin, and more exquisitely refined than either, yet bearing to both of them a stronger affinity, both in the roots of verbs and in the forms of grammar, than could possibly have been produced by accident; so strong, indeed, that no philologer could examine them all three, without believing them to have sprung from some common source."

Sir William Jones, Asiatic Society, Calcutta, 02/02/1786.

The simplest geometric figure one could represent with only four points in a 3D space is the pyramid. Pyramids are the simplest form of orthogonal geometric figures and they are the true representation of our Reality.

If you project a point image on each side of the four vortices of a pyramid you'll obtain eight points altogether. If you consider one corner of a pyramid as the beginning of one cycle you'll have to count seven different corners before you arrive to the original point of departure.

It's this spatial aberration found earlier in our analysis the one responsible for the shrinking of an ideal ten-sub-dimensional space that cannot exist into a shorter eight-sub-dimensional one with an extraordinary nature. Atoms absorb energy due to this symmetry miracle of Reality. It is in the making of a quantum of Reality that space and time suffer from a sudden cancellation allowing energy to occupy the vacuum left by the void. Since those "links" (we saw earlier in our examples) are interrelated by an "object-image" relationship the energy absorbed will mimic the same orbital configuration where it was caught and scattered later on as the cycle resumes and begins a new one.

Aristotle once though that some numbers should have been practically eliminated from the mathematic language. He added **1, 2, 3** and **4** and he came up with **10.** The moment he continued with the application of the same rule he found an error; **6, 7, 8,** and **9** should also add up for the total of **10,** but they represented **30** instead. The universe was obviously designed to a total of 10 numbers and yet he couldn't figure out that **30** also had a meaning on its own.

The only way to properly fit **30** in a series of **10s** was only conceivable by a thought sprung from logic and common sense: **a 3D space.**

The moment you begin to think three-dimensionally and cyclically you begin to see why those early philosophers weren't unable to come up with the right solution.

Plato always believed that the royal way to understanding Reality will flow from the true understanding of the "connection" between numbers and what they represent.

I simply took what they left for me to solve. I wish I could go back and have the chance to meet face-to-face with each and every one of them. It was indeed a marvelous time in story humanity. Without "Particle Accelerators and sophisticated labs to prove their hypothesis; only human intuition and two God-given gifts: Logic and Common Sense.

You saw earlier the mathematical proof explaining why were those two "spatial solutions" cancelled within a "Quantum of Reality." The cancellation obviously had to affect every single aspect of physics whether it was the "coupling constant" or the radius of the orbit followed by an 'electron', the speed of light or the strength of the magnetic field. Everything is interconnected with a Reality that goes beyond our capacity to observe; but not beyond our capacity to understand it. After the cancellation of those "existential instants" came the elimination of the attributes endowed to each one of them; the scale-representing magnitudes and leaving us with one constant and mysterious number **(137).**

I had to grab that number and take it with me "through the looking glass" to see how it looks on the other side. I did just what Alice did many times to figure out the true nature of the red and the white queens. I only followed Lewis Carroll's strategy that's all I did.

You know... Sometimes there is more true hidden in a child story that there is in an entire encyclopedia; isn't it life sometimes pretty ironic?

"Once upon a time there were two Queens; one was the "White Queen" and the other the "Red Queen.". The "White Queen" was always acting on pure **logic.** Her kingdom was a model of **symmetry** and perfection. For every flower there was an **image** and every day the Sun rose from the East and set in the West. She knew a game that would cancel those sets of numbers in the 10th

scale and supply **energy** with **a wave** configuration...." The symbol posted in front of the "White Queen's castle" read like this:

"...0-1234-5-6789-0-1234-5-6789-0-..."

"Then there was the "Red Queen." She didn't have the perfection the "White Queen" used to flatter herself but she was the model for **simplicity** and honesty. The "Red Queen" was the incarnation of **common sense.** Her kingdom was covered by **space.** She had **three dimensions** or "gardens" which she gave them the names of **"length", "height" and "width".**

The "Red Queens's" symbol was an "**Orthogonal** symbol" she called it a **"Pyramid."**

She also used to flatter herself saying that only in her kingdom you could build a geometrical figure in **3d space** and she also knew a secret game to turn energy into four "object-points" and their corresponding "Image-points." She called them the nickname of 'electrons'. She knows they're not actually real all the time but for sake of the game four points with their images was more than enough.

One day Alice needed the help of both majesties and both refused to help her. Alice had an idea. She took both "through the looking glass" and discovered their true nature and only then Alice was able to convince both of them to work out a common solution to incorporate both kingdoms into just one. Alice already knew the secret "key" that open both gates: it was 137.

Her "magic key" was the bridge between **logic** and **common sense;** between an impossible "**10** subdimensional Space" represented by **10** real numbers and a realistic "**3-dimensional one**" with a cyclical configuration.

Alice was finally happy. The "magic key" worked wonders it was then and only then that she finally understood that her **"Mirror"** was the only viable **way** to look beyond the appearance showed by the numbers. After all it was simply natural; every time she wanted to know whether a new dress or a new hat fit nicely she always went to see herself on the mirror... Why would this time such a simple thing have to be different?

Alice finally understood that you need two ways always: the way of **logic** and the way of **common sense**."

I found an interesting article written and posted in the Web by an Italian psychologist named Giorgio Piacenza Cabrera, after trying to reach him in many occasions without success I took the liberty to quote some of his words in regards to the mystery behind **137:**

"...One of these hints seems to be the recently discovered fact that number 1/137 multiplied by various powers of 10 recurs every time

we cancel the dimensions and form non dimensional numbers using atomic constants..."

But Alice already knew that since **1/137** was her **"Magic Key"** to 'open the gate' between those two castles and finally solve the confrontation between the concept of **10s** and the nature of **"cycles"** with an objective **3D** space composed by the "Red Queen's gardens".

Let's read another one:

"When the "fine" or closely positioned double spectral lines of the hydrogen spectrum were observed by Sommerfeld in 1915, he decided to broaden Bohr's atomic model by allowing for elliptical orbits and how speed affected mass. While trying to find the frequencies of the spectral lines, he came up with "alpha" which can be understood as a fundamental constant and –because of being dimensionless- can also be understood as a ratio. The units of this number are Coulombs, meters per second and Joules per second but they *cancel out*, leaving a dimensionless ratio, *applicable in any measurement system*, a ratio which we could think of as independent from how our minds work or quantify..."

Oh! He must be referring to the "magic trick" used by the "White Queen". She knew her ways to cancel each other those number she had posted in front of her castle, remember the story?

Of course the "white queen's magic trick" was... "Applicable in any measurement system" because every single measurement system in our universe must respect the compromising solution reached by 'Alice' between both "Queens."

That's why those serious men of science couldn't find any unit to explain this unexpected result. They were left with a mysterious number, a mathematical 'creature' with a name but without a last name; 'someone' without a family, a country of origin not even a residential address.

All they know was that without this "someone" nothing could be calculated and nothing has a meaning. Alice already knew that things only acquire a meaning after they are taken 'through a looking glass." When Alice took both "queens" with her and put them in front of her mirror she could understand what those two "queens" actually were in the inside.

Let's continue reading his words:

"...In 1935, he (Max Born) delivered a lecture at the Indian Scientific Association. The lecture was entitled, "The Mysterious Number 137." Born said that –significantly- alpha derived from a combination of the electron's charge, the speed of light and Planck's Constant and, since those times, the "arbitrary" value of 137 has been understood by the majority of scientists as a matter of chance,

while a few others like Paul Davies PhD, the number can be associated with an organizing o selective power in the Universe..."

Congratulations Dr. Davies! You may not knew in details "Alice's story in the Wonderland" as we know but you had the intuition to see that whatever was 'producing' that constant obeyed to a higher order of thing...

Let's see another one:

"It is said that Plato considered (in Pythagorean fashion) that the study of numbers 'without bodies' was the most important of all sciences. Perhaps Plato's basic intuitions and the dream of developing a science primarily based on reason still holds in spite of the emphasis recently give to empiricism. Maybe in today's scientific world, the ratio 1/137 is the look of a feel that cannot be extricated from a reality which includes interiority. Perhaps 1/137 serves to remind us of a higher call par excellence. Could the number only be a meaningless coincidence, a necessary but neutral mathematical consequence of reducing the constants to their common elements when canceling their dimensions or, perhaps, it might as well be the essence underlying the constants, thus "la Crème de la Crème" of our physical knowledge?

I think you all know all answer to the last part. As for Plato's vision I could only say that he was one of my favorite philosophers in the entire history of humanity. He was right back then when there wasn't a "particle Accelerator" nearby to confront his thoughts. As the author of the article said, "Plato's intuition" turned out to be the correct one even in the absence of a required **3D** and cyclical 'State of Mind' to understand that even numbers have to adapt themselves to a compromising solution in search for the correct model capable of representing our Reality... How could he possibly?

Last one:

"Cornwell points out that Einstein had asked whether "God had any choice when he established the precise and seemingly random constants that make up the measurable universe," and then call attention to the mystery of number 137.03604 by saying that "A remarkable illustration of fine tuning is this: when mathematicians, in search of a fundamental equation, square the charge of the electron and divide it by the speed of light times Planck's constant, the dimensions (mass, time and distance) cancel out, yielding the number known as alpha (also known as the fine structure constant) which is approximately 1/137. What is the reason for this peculiar number which appears to underpin the whole of nature? Some scientists believe that the solution could lead, ultimately, to *a grand theory of everything.* "

I believe it too ☻

I hope you had enjoyed my colorful analogies and I hope that no one would take this the wrong way. I want to say that I've always felt the outmost respect for every single physicist in the search for the final truth whoever he or she was. However I also believe that no one (Except for God itself) has the right to claim the monopoly of the truth. We were all blessed by its gifts and whether one could speak the language of mathematics or not that doesn't make him or her less capable to understand the universe shared by everyone of us.

137 is everywhere we look at; from the remote corner of the universe to the unreachable distances of the 'Quantum World'; from the very essence of the human spirit to the miracle of life and its own destiny...Not even the price of this book was safe from its devastating influence.

The value of 'Constant Alpha' in all four "Dimensional Existence"

The only comfort I could offer to you is that there must be a whole bunch of people out there worrying for their own "**137**'s"... Allow me to explain.

137 is only a dimensionless multi-task and multi-purpose constant in the confines of our existence alone. Our case is somewhat similar to the other three co-existent Realities however the number is not the same.

But first let me make a "laundry list" of what we know now:

1) Symmetry cannot be violated under any circumstance; **11 and 16.**

2) Universal Perturbation Factor could only take the numbers **(+1)** and/or **(-1).**

3) Energy-mass conservation can only take the values of **11** and/or **16.**

First Universe where Alpha = 137:

Let's see how our **317** fulfill those basic requirements (Conditions).

Symmetry and conservation of mass/energy:

(137) [+1+3+7=11]

(682) [+6+8+2=16] * 862 is the image of 137

Perturbation:

(000) (137) P= {-1-3+3} Pt = (-1)

(555) (682) P= {-1-3+3} Pt = (-1) *Image= (+1) as we saw earlier.

Second Universe where Alpha= 128:

(128) [+1+2+8=11]

(673) [+6+7+3= 16]

Perturbation:

(000) (128) P= {-1-2+2} Pt = (+1) *Image= (-1)

(555) (673) P= {+1-2+2} Pt = (+1)

Third Universe where Alpha= 146

(146) [+1+4+6= 11]

(691) [+6+9+1= 16]

Perturbation:

(000) (146) P= {-1-4+4} Pt = (-1)

(555) (691) P= {-1-4+4} Pt = (-1) *Image= (+1)

Forth Universe where Alpha= 236:

(236) [+2+3+6= 11]

(781) [+7+8+1= 16]

Perturbation:

(000) (236) P= {-2-3+4} Pt = (-1)

(555) (781) P= {-2-3+4} Pt = (-1)

These are the four possible universes occupying the same space but in separate time lapses.

Below you'll find some of the 'Impossible Universes' that cannot be stable under the rules that I listed above:

First impossible universe where Alpha is 425:

(425) [+4+2+5= 11] Ok!

(970) [+9+7+0= 16] Ok!

Perturbation:

(000) (425) P= {-4-2+5} Pt = (-1)

(555) (970) P= {-4-2+5} Pt = (-1)

What makes this universe impossible is the fact that one of the coordinates is always 0 or 5 and that makes space a plane instead of 3D. Atoms have anti-bonding axes where molecules can't form.

Second impossible universe where Alpha is 344:

(344) [+3+4+4= 11] Ok!

(899) [+8+9+9= 26] Wrong!

Third impossible universe where Alpha is 353:

(353) [+3+5+3= 11] Ok!

(808) [+8+0+8= 16] Ok!

Perturbation:

(000) (353) and (555) (808) are planes and not 3d space. After cancellation there will be no sets of coordinates for point mass in 3d space.

Fourth impossible Universe where Alpha is 191:

(191) [+1+9+1= 11] Ok!

(646) [+6+4+6= 16] Ok!

Perturbation:

(000) (191) P= {-1+1-1} Pt= (-1)

(555) (191) P= {+4-4-4} Pt= (+4) Wrong!

These are some of the impossible universes or forbidden existences that are not permitted under the three conservation laws; namely mass/energy, symmetry and perturbation.

There is a note that I'd like to add to this analysis and that has to do with the number **11** and **16**.

The meaning of **11** gives us the notion of "objects" allowed for orbital; as in the case for 'electrons' there can only be two per 'subshells'.

The meaning of **16** represents the notion of **'Spin'**. We know that every "particle" must have its own "Image-particle" just as every **(+)** must have its own **(-)**. The meaning of positive only becomes legitimate if there is a referential image or its corresponding negative.

As we saw during the determination of the hidden meaning of **137** in *our Universe* the fractional or decimal portion of the whole constant is necessary to verify that it does comply with the 3D configuration or **(-7)**.

The need of a dimensionless constant arises from the fact that the only number capable of representing true symmetry is **1.**

Where 1 divided by 1 equals 1, where 1 is its own squared and the square root of 1 is also 1.

I can only guess that those 'possible universes' that I mentioned above will have also a similar fraction giving them the legitimacy required to exist in the first place. Obviously we are not allowed by the nature of Time and Reality to make the necessary "crossover" to determine those numbers...For what I can guess the answer to this puzzle maybe hidden in the disturbances observed inside the "Bermuda Triangle."

The Mysterious meaning of π (Pi) was also answered:

This well known mathematical constant has been around for centuries...

3.14159 26535 89793 23846 26433 83279 50288 41971 69399 37510.

It is defined as the circumference of a circle with diameter 1.
The following quote (extracted from Wikipedia) gives us an idea of how important the notion of the origin of this constant could be for the development of a "Unified Theory."

"...Pi's universality is not limited to mathematics. Indeed, various formulas in physics, such as Heisenberg's uncertainty principle, and constants such as the cosmological constant bear the constant pi. The presence of pi in physical principles, laws and formulas can have very simple explanations. For example, Coulomb's law describing the inverse square proportionality of the magnitude of the electrostatic force between two electric charges and their distance, states that, in Si units..." (End of Quote)
Let's bring back the "little number" back into this page and "cut it out" in pieces, transform it into "object-only patterns" and apply my 'Mirror Analysis' and see what happens to this "long fellow."
3.14159 26535 89793 23846 26433 83279 50288 41971 69399 37510
A) First thing I'm going to do it is to cut it in equal number of pieces:
(14159) (26535) (89793) (23846) (26433) (83279) (50288) (41971) (69399) (37510) Good!

B) Now we're going to do exactly the same trick I learned from the analysis of Cash 3:

(14104) (21030) (34243) (23341) (21433) (33224) (00233) (41421) (14344) (32010)

I bet you see what I see...

C) Let's match them by symmetry:

(14159) (41471) P= {0-1-3-30} Pt = (-7)

(21030) (32010) P= {-3+100+2} Pt = (0)

(34243) (33224) P= {0+10+2-1} Pt = (+2) = (-7)!

(23341) (21433) P= {-1+3-10-1} Pt = (0)

(Continuing with the 'Mirror Analysis' of Pi)

(00233) (32010) P= {0+1+2-10} Pt = (+2) = (-7)!

So! Is this new mathematical approach telling you something about the "hidden secret" behind one of the oldest constants of the universe? I bet it is.

The symmetry results for the Pi constant mimics the same we saw in the case of 317!

Gosh! I wish Dr. Feynman was here now! You know...In a sense he's been here all the time watching with his smile as I went through these numbers.

I'd like to dedicate this discovery to honor his memory and his long and fruitful life spent in the quest for the "Unified Theory."

This is what he had to say with respect to the relationship between Alpha and Pi:

"...It has been a <u>mystery</u> ever since it was discovered more than fifty years ago, and all good theoretical physicists put this number up on their wall and worry about it.) Immediately you would like to know where this number for a coupling comes from: is it related to pi or perhaps to the base of natural logarithms? Nobody knows. It's one of the greatest damn mysteries of physics: a magic number that comes to us with no understanding by man..."

<u>Now the mathematical proof:</u>

α Alpha = 137.03604...

Π (Pi) = 03.141592...

(0314) (3604) P= {+10+3-4} Pt = (0)!

(1542) (1370) P= {+1+1-2+1} Pt = (+1)! <small>Both were the same!</small>

The symmetrical parallel between both universal constants equal +1; which means that they both have been one and the same all this time. The reason that allowed them to remained secretly apart was that Alpha was the "Magic Key" for numbers multiples of 10 or exponential of 10. Pi on the other side was the shorter version and the one referring to the smallest unit = 1.

The fact that they don't look very much alike is because we must account for the changes introduced to them by the consequences of Perturbations in between both scales.

Conclusions:

Many!

a) One is definitely that (-7) or (2) [As its opposite intrinsic value] means the presence of "Object" and "Image" in that orbital.

b) Adding 1+3+7= 11 *It could be taken as "eleven" but it also means "1 plus 1" which equals 2!

c) Replacing 7 for its "object" in the expression 137 we will have: 1+3+2 = 6! *That is the "Image" of number 1.

d) Replacing all the "objects" for their "Image" in 137: 6+8+2= 16! Representing 1 the "Object" and 6 its corresponding "Image".

CHAPTER X

"End of the cycle"

It is my sincere hope you will find this little book both useful and entertaining. Five years ago I felt that I had something to say and I felt I did it using the best of my abilities. Today I'm convinced that I found the evidence I was looking for back then when I self-published my first book with X-libris under the title "Warning Scientific Discretion Advised."

Who knows? Everything may well be just a bunch of nonsense with no scientific basis to where one could start building a solid theory. For me those patterns are more than just a convenient arrangement of disconnected numbers with no meaning and no reason behind its futile existence.

Those numbers have spoken to me in their own encrypted "language" and they've convinced me about the existence of an incredible force that rules everything and everyone in ways one could hardly imagine.

I can't deny that at moments I was seriously worried about the unusual effect those patterns were affecting my daily life. I thought I was "creating' them in my mind instead of interpreting a hidden meaning that those numbers at least apparently represented. I turned to science for help and one day I found a concept that gave me the needed encouragement that have helped me to keep on with my search until I found myself where I am now.

The concept was coined "Synchronicity" and his author was a Swiss psychologist named Carl Gustav Jung.

Dr. Jung used to refer to it as a "temporarily coincident occurrences of "acasual events". Although the principle was discovered in 1920 it wasn't actually published until 1951, guess who co-author the scientific paper book along with Dr. Jung?

You're darn right! Dr. Wolfgang Pauli himself!

Both agreed that what sometimes is referred as 'archetypes' or 'collective unconsciousness' was the 'governing dynamic' that underlies the entire human experience and has an unavoidable influence in its history; acting within the realms of the spiritual, social emotional and the psychological facets of human beings.

That was a wonderful day in my life. I can say that that was the day I knew I was not crazy or at least not yet...Anyway!

Dr. Jung's work implied that the "Law of Probabilities" seemed to become vulnerable and even under the weight of multiple coincidences violated in some instances. Perhaps a more skeptical type personality would say that the "Laws of Probability" also implies certain degree of 'probability' for a violation of its own statue as universal law ☻

I disagree with such hypothesis even before anyone dared to imply such idiocy. A law is a law because it cannot be violated. If the 'Law of Probabilities" happens to be susceptible to be violated in any sense (As I have already proven with enough evidences in the case of the "Cash 3" game) then the very essence of its statue as a law of nature will become undeniably questionable.

Those patterns I discovered, classified and analyzed from a popular game of luck supplied my mind with the necessary intuitive abilities to unveil the connection hidden beyond images and objects; between links of the same 'power series'. It wasn't a coincidence either that Dr. Jung's favorite quote on Synchronicity was from "Through the Looking Glass" by Lewis Carroll when the white queen said to Alice: **"It's a poor sort of memory that only works backwards."**

I always knew that the answer to our questions were to be found beyond the "Looking Glass" of Alice's wonderland; I just never knew that I was going to find it the same day I decided to crossover to the other side of the same mirror carrying on with me a bunch of numbers over my shoulders.

Five years ago I used a child story in my first book. I tried to represent the sense of absurdity and incompleteness embedded in the nature of Quantum Mechanics along with its dangerous and paradoxical message in a child story by Hans Christian Andersen titled "The Emperor's Magic clothes". I made a parallelism between the so-called "wave-function of probabilities" and the interpretations derived from the acceptance of the "Standard Model of Particles" with the 'invisible and unrealistic' nature of the fabric used by a crook to tailor the Emperor's clothes.

I surely hope that I was able to deliver the message back then with the same level of clarity and simplicity as I did today using the story of "Alice in Wonderland" to unveil the mystery behind the powerful meaning of **137.**

To me the human genius and creativity is endless as it is the amount of space in the universe. We were given a set of precious gifts but we must use them wisely and that means having balance in finding a suitable compromise in the face of a confrontation. To live strictly by the rules of logic is as wrong as it is to live by those defining Common Sense alone. The first would make us automatons in our own way and the second would be lacking the laws and principles that rule the universe ultimately leading humanity into chaos and anarchy.

Unfortunately there isn't a "magic key" like **137** to bring the needed balance into the human spirit as it was the case with the orbits of 'electrons' or the magnitude of the 'speed of light'.

Atoms don't have 'free will' and for 'they' the guarantee of having **137** would suffice their existence. We must find our own **137** too if we really wish to survive as a race.

We must find a sense of balance and equilibrium in our lives as well as within the structure of the complexity built inside the human society. We must understand that there is no sense for good or bad without a system of reference and this system had always been the same since the beginnings of the beginnings: **"It"** is called God.

There is no bright without dark and there is no love without hate. Those human emotions are as perfectly natural and legitimate as there are the magnetic field and its counterpart; the electric field, matter and anti-matter or sound and silence. Finding a perfect balance between them is the best way to appreciate God's gifts and to respect its teachings.

One last calculation and I promise no more numbers:

(000) (111) P= {-1-1-1} Pt = (-3)

(000) (611) P= {+4-1-1} Pt = (+2)

(000) (161) P= {-1+4-1} Pt = (+2)

(000) (116) P= {-1-1+4} Pt = (+2)

(000) (661) P= {+4+4-1} Pt = (+7)

(000) (616) P= {+4-1+4} Pt = (+7)

(000) (166) P= {-1+4+4} Pt = (+7)

(000) (666) P= {+4+4+4} Pt = (+12)

Symmetrical cancellations:

(111) (666) P= {000} Pt = (0)!

(166) (661) P= {000} Pt = (0)

(661) (166) P= {000} Pt = (0)

(161) (616) P= {-5+5-5} Pt = (-5) = (0)

Adding all the Perturbations:

(-3) (+2) (+2) (+2) (+7) (+7) (+7) (+12) = (36)

Let's transform (+7) in their equivalence of "object":

(-3) (+2) (+2) (+2) (-2) (-2) (-2) (+12) = (+9) or (-4) or $\frac{1}{4}$.

4X9= 36. Also 4X90° = 360°

And if we 'mirror' both:

(003) (600) P= {-300} Pt = (-3) = (+8) Eight are the numbers of actual materializations of energy ("Objects" plus "images") in a 3D space within the time equivalent to a wavelength of light = 1 "Quantum of Reality."

What is the hidden message behind the true 'face' of 137?

Perhaps one day both mathematicians and physicists will agree in that my results and their final conclusion (-7) and (-1) were much more than a simple numerical coincidence.

Before I conclude with my "PROOF" I had a last surprise...

An interesting calculation that you'll love it as much as I did:

If we take -7 as 1/7 according to Algebraic rules, then:

$1/7 = 0.1428571$ *Applying the 'Mirror'...

a) (142) (857) P= {+4-1+4} Pt= 7. Changing (857) for (302)

b) (142) (302) P= {-1+4-1} Pt = 2. Changing (142) by (697)

c) (697) (857) P= {-1+4-1} Pt = 2. Changing both...

d) (697) (302) P= {-4-1+4} Pt = -1. (Also Irrational!)

Since we already know 2= -7 also, then we have proved that the "Mirror Analysis" applied to the numerical representation of 1/7 coincides with the same results obtained during the same analysis on 137!

Isn't it something?

Such a remarkable display of symmetry and beauty could only come from God.

ONE LAST WORD

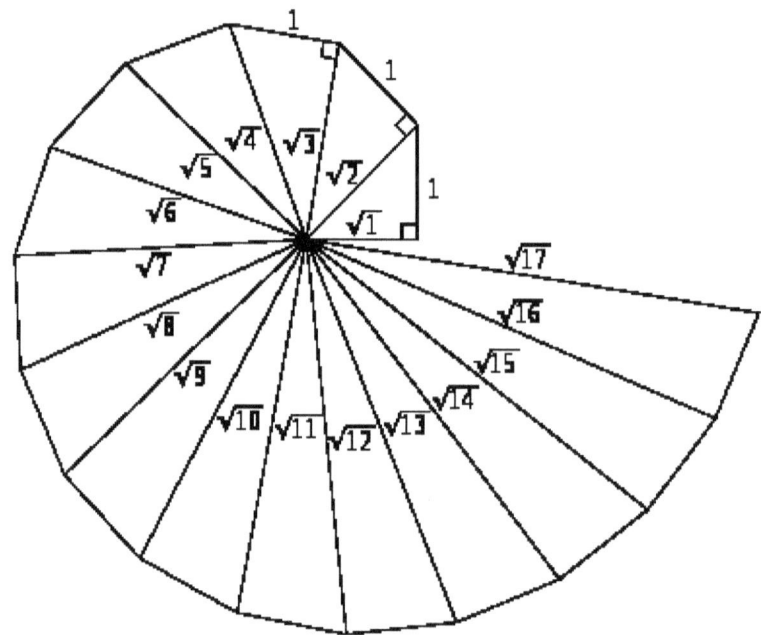

"The golden ratio has fascinated Western intellectuals of diverse interests for at least 2,400 years: some of the greatest mathematical minds of all ages, from Pythagoras and Euclid in ancient Greece, through the medieval Italian mathematician Leonardo of Pisa and the Renaissance astronomer Johannes Keppler, to present-day scientific figures such as Oxford physicist Roger Penrose, have spent endless hours over this simple ratio and its properties. But the fascination with the Golden Ratio is not confined just to mathematicians. Biologists, artists, musicians, historians, architects, psychologists, and even mystics have pondered and debated the basis of its ubiquity and appeal. **In fact, it is probably fair to say that the Golden Ratio has inspired thinkers of all disciplines like no other number in the history of mathematics".** (Mario Livio, *the Golden Ratio: The Story of Phi, the World's Most Astonishing Number)*

This is complete number as it appears in Wikipedia:

1.6180 33988 74989 48482

As you can observed I separated the number in 4 groups of 5 digits each. I'm going to "Mirror-Analyze" the famous 'Golden Ratio' and discover its most 'intimate secrets':

(16180) (33988) let's bring them down to their 'Object Configuration':

(11130) (33433) P= {-30-3-2-2} Pt= (-10) or (+5)

The same result could be obtained by converting the first 'link' into its "Image"

(16180) = (61635)

(61635) (33988) P= {+20-3+3-2} Pt= (0) Adding 5 for compensation 0+5=5!

This means that the secret will be found in the next two 'links':

(74989) (48482) P= {+50-5-4+5} Pt= (+1) plus half-cycle (-1)!!!!

Since we have half-cycle from the first symmetrical operation (5) it changes the sign of +1 into (-1). It is exactly the same result obtained while operating my "Mirror Analysis" on the square root of 2. Both are considered 'Irrational numbers'.

My conclusion is pretty obvious: "Irrational numbers are representative of an inherent Perturbation of (-1). Their decimal and fractional segments cancel each other out to produce (-1) having part of their 'numerical representation' half-cycle ahead in TIME in 3D space with respect to the rest of the number."

$\sqrt{2} = 1.4142135$

(1414) (2135) P= {+20+1-4} Pt= (-1)!!!

You see? The mystery behind Irrational numbers was unveiled when we take them through the looking glass and look at their own meaning.

The "Spiral of Theodorus" also known as "Einstein Spiral" was originally designed by contiguous right triangles.

"Plato, tutored by Theodorus, questioned why Theodorus stopped at √17. The reason is commonly believed to be that the √17 hypotenuse belongs to the last triangle that does not overlap the figure."

The $\sqrt{17} = 4.1231056$

(041) (231) P= {-1+1-1} (-1)

(231) (056) P= {+1-2-4} (-5) or (0)!

The square root of 17 implies an inherent Perturbation equal to (-1)!

I know Mathematicians would love this blog! However physicists will love the next one:

(In a previous blog of mine titled **"THE SECRET BEHIND IRRATIONAL NUMBERS: THE GOLDEN RATIO"** (Published on June/2010 on www.integrallife.com) I proved that classical irrational numbers like $\sqrt{2}$, $\sqrt{17}$ and 'The golden Ratio" among others had a single peculiarity in common: After operating the "Mirror Analysis" on every one of them I obtained the same result Pt = (-1)! >>>Total Perturbation was always (-1).

You may ask yourselves, what does it mean after all? The question is perfectly legitimate since the best Mathematicians of couldn't answer it to you either. The response lies within the nature of our Reality.

The physical interpretation of why certain numbers and mathematical functions tend to Infinite and must be "Renormalized" into a "shorter version of its own" responds precisely to this issue. I have discovered that part of the irrational number of function is "referenced ahead in time by half-cycle" and that's the WHY we always find a net perturbation of (-1) in almost all important mathematical constants used in Physics; namely "Alpha", "Pi", "The Golden Ratio", etc.

Since a "Quantum of Reality" cannot be broken in pieces the solution for the square of irrational numbers and those constants that I mentioned earlier will extend their fractional part toward infinite. This doesn't mean that at some point in time there will be a final exact solution for them, on the contrary the solution is to be found in the 'links' used during my "Mirror Analysis". Sometimes I needed only 2 (two) 'links' to obtain Pt = -1 or Pt = -2 however as in the case with "Pi" I had to relay in a complete analysis of all 50 (fifty) decimals to verify which ones cancel each other out and which ones stay after all the proper operations are done with.

It all comes to the basic point: Reference! Imagine that you are watching the Olympic Games and there is an athletic competition where runners will participate in the 100 meters match. Then suddenly one of the runners simply walks ahead half of the way to the end and decided to start from there... You could imagine what will happen to that runner afterwards but my point is that the match would have probably be suspended until every runner gets ready behind the same line having all the same distance to cover for the final prize. Reality is even stricter than those judges at the Olympic Games. "Alpha constant" is the mathematical solution to operate on a physical phenomenon happening within the time-constraints of a "Quantum of Reality".

That's why as well as "Pi" and the other irrational numbers and functions, "Alpha" too produces a net perturbation of (-1) when the "Mirror" was placed in between 'links' properly selected inside the long fractional number that represents its mathematical value.

"PLANK'S CONSTANT WAS "17" THROUGH THE LOOKING GLASS"

$6.62606896(33) \times 10^{-34}$ "Joules per second"

$1.054571628(53) \times 10^{-27}$ "Ergs per second"

Let's begin with the first from above:

(6626) (6896) (3) *"0" can be skipped by Reality and "3" Repeats itself toward infinite.

Let's convert this number into its "Object configuration" first:

(1121) (1134) (3) let's add all those numbers: **(5) (9) (3) = (17)!!!**

Since there were no digit (5) in the decimal we could assume that the whole number was referenced to the cycle's origin = 0! However watch for the next number...

Now let's continue with the second number below:

(1054) (5716) (2853)

Let's do it including the **"0"** first while converting it to the 'object configuration':

(1004) (0211) (2303) >>> (5) (4) (8) = (17)!!!

You think I was 'unfair' because I left the **"0"** in the first 'link' right? OK! Let's take it out!

(1040) (2112) (3030) >>> (5) (6) (6) = (17)!!!

Now what?

Did you realize that the 'whole part' of the number in both numerical representation of the Plank's Constant was related under the "Object" to "Image" correspondence?

I've said that "6" was the image of "1" (To obtain a number's image just add 5 to it, it's so simple!)

Now! What do you think is the 'UNIVERSE' telling us here? I tell you what:

"IT" is telling us that decimal points and zeros were all but IRRELEVANT and that Reality has never really cared much about it...

We care so much about them because we exist in the form of 'matter' in a UNIVERSE that is PURE ENERGY.

***

I decided to leave for the conclusion of this book a quick synthesis of what I wrote more than five years ago in a book that had to be "dressed out" as fiction for obvious reasons:

"...If I were in those Alien's shoes (assuming they have foot like we do to begin with) I would be more interested in the way humans see Reality.

Shouldn't Reality represent the solid basis upon which our civilization built its magnificent edifice of knowledge? Hasn't been always Reality the theoretical support of human identity, the mirror where our most secret beliefs, from Truth to Common sense have always found their own image?

The problem is that our Reality and the science supposed to describe it have been in deep crisis for a while now, and I'm not referring to the historical confrontation between science and

religion, facts and faith... theorems and commandments. That one, I'm afraid, has been going on for centuries with very little hope to end anytime soon..."

"...Only this time, science finds itself in trouble with Reality, which also means that it is in trouble with itself.

"If we were to assume common sense as a criterion for the truth or falsity of a physical theory, we would have to reject the two most basic theories of modern science, the theory of relativity and the quantum theory." –From "The Concept of Science".

Could you tell me what's wrong with this picture?" (End of quote from "Warning Scientific Discretion Advised" M. De Zayas 2005)

There is not a day that goes by without a man, a, women or even a child having to die or see their families and homes destroyed "in the name of God." If God had something to teach us, something at all it was how the search for a common ground, the fight for a mutually beneficial compromise will be the royal way that will lead us to find our own destiny.

What a better example of wisdom than to be able to compromise between two different dominium: one owing pure logic, cycles with ten numbers and energy in the form of waves; the second representing common sense, 3D space and an orthogonal figure (pyramid), representing its simplicity, beauty and perfection?

"It" did it with grace, art and harmony. IT knew that something had to be sacrificed for 'the greater good' so IT took matters into IT's "own hands" and created a constant. We thought there were two when it was just one because we never really knew what the purpose of the constant was. Alpha or Pi were designed and endowed with universal powers to unify both dominium.

It was that using **137** men could unveil the mysteries of the atom and the cosmos but couldn't understand the true mystery behind it. **137** was God's guarantee for unification, for coexistence between two worlds so different from each other and unique in their own nature.

Humanity must find its own rules and ways to coexist. We are connected in ways we cannot understand. The human race must find their own **137;** a perfect solution for an imperfect world.

We must understand that we were the condition for the great unification. A material universe was the miracle that made possible such communion and we must cherish every moment of our life doing exactly what IT did and taught us to do.

Just as the "White Queen" saw some of her energy and numbers cancelling each other out and having to share her kingdom with another. On the other hand, the "Red Queen" had to give up part of her freedom and share it with logic, principles and laws that she didn't need before. Hers was the space and the three dimensions she lived simple and happy until Alice came for help.

One last point I'd like to make before this book finds its own conclusion.

666 has been since the very beginning the dark face of **111;** not the "evil" but the "image" of 'IT'.

There cannot be love without hate, or freedom without slavery or happiness without sadness. For centuries we had erroneously judged the nature of **666** using it as a generating source of evil in the world. Nothing could be farther from the truth. The so-called "dark matter" is as real as we are on this side of the universe. "Objects" cannot exist without their own "images" just as **111** must rely on **666** for eternal coexistence.

That was the reason that I decided to place the "Yin and Yang" symbol in this book. It represents the two-face Reality and its eternal nature. If something you've learned from this little book was that sometimes we must face "objects" against their own "images" to really understand who they really are in the inside. I did it with numbers and it worked by showing us the answer to secret constants that our best minds have been searching unsuccessfully for decades. Humans are no different than numbers in that sense. We must see ourselves in the image of God. We must understand who we really are and what we really look for in this life but the only way to achieve such difficult task is by accepting God's examples and make it our own.

By 'compromising' and searching for common goals that will eventually benefit both parties as well. Killing in the name of God is the same as negating our own Reality and therefore God's creation itself. We could spend the rest of our brief existence arguing who's the "good guy" and who is the "bad guy" because we are in charge of the final position of the 'mirror' and we do as we see it better. We have been deciding who the "object" and who the "image" was so we could claim to be right and legitimate before God and that's wrong and unfair.

God is and have always been the sole 'Mirror" of our souls and the unique judge of our deeds. Those who see themselves in its mirror will find their way and the true meaning of Reality.

God bless you.

www.ingramcontent.com/pod-product-compliance
Lightning Source LLC
Chambersburg PA
CBHW051519170526
45165CB00002B/534